Springer Theses

Recognizing Outstanding Ph.D. Research

For further volumes:
http://www.springer.com/series/8790

Aims and Scope

The series "Springer Theses" brings together a selection of the very best Ph.D. theses from around the world and across the physical sciences. Nominated and endorsed by two recognized specialists, each published volume has been selected for its scientific excellence and the high impact of its contents for the pertinent field of research. For greater accessibility to non-specialists, the published versions include an extended introduction, as well as a foreword by the student's supervisor explaining the special relevance of the work for the field. As a whole, the series will provide a valuable resource both for newcomers to the research fields described, and for other scientists seeking detailed background information on special questions. Finally, it provides an accredited documentation of the valuable contributions made by today's younger generation of scientists.

Theses are accepted into the series by invited nomination only and must fulfill all of the following criteria

- They must be written in good English.
- The topic should fall within the confines of Chemistry, Physics, Earth Sciences, Engineering and related interdisciplinary fields such as Materials, Nanoscience, Chemical Engineering, Complex Systems and Biophysics.
- The work reported in the thesis must represent a significant scientific advance.
- If the thesis includes previously published material, permission to reproduce this must be gained from the respective copyright holder.
- They must have been examined and passed during the 12 months prior to nomination.
- Each thesis should include a foreword by the supervisor outlining the significance of its content.
- The theses should have a clearly defined structure including an introduction accessible to scientists not expert in that particular field.

Jonelle Harvey

Modelling the Dissociation Dynamics and Threshold Photoelectron Spectra of Small Halogenated Molecules

Doctoral Thesis accepted by
the University of Birmingham, UK

 Springer

Author
Dr. Jonelle Harvey
School of Chemistry
University of Birmingham
Birmingham
UK

Supervisor
Prof. Richard P. Tuckett
School of Chemistry
University of Birmingham
Birmingham
UK

ISSN 2190-5053 ISSN 2190-5061 (electronic)
ISBN 978-3-319-02975-7 ISBN 978-3-319-02976-4 (eBook)
DOI 10.1007/978-3-319-02976-4
Springer Cham Heidelberg New York Dordrecht London

Library of Congress Control Number: 2013953192

© Springer International Publishing Switzerland 2014
This work is subject to copyright. All rights are reserved by the Publisher, whether the whole or part of the material is concerned, specifically the rights of translation, reprinting, reuse of illustrations, recitation, broadcasting, reproduction on microfilms or in any other physical way, and transmission or information storage and retrieval, electronic adaptation, computer software, or by similar or dissimilar methodology now known or hereafter developed. Exempted from this legal reservation are brief excerpts in connection with reviews or scholarly analysis or material supplied specifically for the purpose of being entered and executed on a computer system, for exclusive use by the purchaser of the work. Duplication of this publication or parts thereof is permitted only under the provisions of the Copyright Law of the Publisher's location, in its current version, and permission for use must always be obtained from Springer. Permissions for use may be obtained through RightsLink at the Copyright Clearance Center. Violations are liable to prosecution under the respective Copyright Law.
The use of general descriptive names, registered names, trademarks, service marks, etc. in this publication does not imply, even in the absence of a specific statement, that such names are exempt from the relevant protective laws and regulations and therefore free for general use.
While the advice and information in this book are believed to be true and accurate at the date of publication, neither the authors nor the editors nor the publisher can accept any legal responsibility for any errors or omissions that may be made. The publisher makes no warranty, express or implied, with respect to the material contained herein.

Printed on acid-free paper

Springer is part of Springer Science+Business Media (www.springer.com)

Parts of this thesis have been published in the following journals:

[1] J. Harvey, R. P. Tuckett and A. Bodi, '*A Halomethane Thermochemical Network from iPEPICO Experiments and Quantum Chemical Calculations*'., Journal of Physical Chemistry A, (2012), volume **116**, issue 39, pages 9696–9705. Reproduced with permission.

[2] J. Harvey, A. Bodi, R. P. Tuckett and B. Sztáray, '*Dissociation dynamics of fluorinated ethene cations: from time bombs on a molecular level to double-regime dissociators*'., Physical Chemistry Chemical Physics, (2012), volume **14**, pages 3935–3948. Reproduced with permission.

[3] J. Harvey, Patrick Hemberger, A. Bodi and R. P. Tuckett, '*Vibrational and electronic excitations in fluorinated ethene cations from the ground up*', Journal of Chemical Physics, (2013), volume **138**, pages 124301–124313. Reproduced with permission.

The data included in Appendices A–C are to be published as forthcoming journal articles.

Supervisor's Foreword

The research described in Jonelle Harvey's thesis awarded for the degree of Doctor of Philosophy involves two experimental aspects of the imaging photoelectron photoion coincidence (iPEPICO) apparatus which is stationed on the X04DB vacuum-ultraviolet beamline of the Swiss Light Source (SLS). This is a third-generation synchrotron source which has been operative since 2001, and is located at the Paul Scherrer Institute, CH 5232 Villigen, Switzerland. Fundamental properties (thermodynamic, spectroscopic and dynamic) of the cations of polyatomic molecules and their daughter ions have been investigated at high resolution, *ca.* 0.001 eV. This thesis has already led to three major papers being published in the international peer-reviewed literature. Further two to three papers will result from work published in the Appendices which is currently being analysed.

The fast and slow dissociation dynamics of halogenated methanes and fluorinated ethenes have been investigated using threshold photoelectron photoion coincidence (TPEPICO) techniques. Rate constants and very accurate appearance energies at 0 K for the formation of subsequent daughter ions have been determined. The latter values have been used in conjunction with a vast array of *ab initio* calculations to derive updated enthalpies of formation of polyatomic molecules, their parent and daughter cations at both 0 and 298 K.

The valence threshold photoelectron spectra of four fluorinated ethenes have been recorded. The spectra have been analysed using Franck-Condon simulations to model the vibrational structure and assign the spectra, sometimes revising previous assignments in the literature. The potential energy surfaces of the ground and excited electronic states of $C_2H_3F^+$ have been explored to uncover their various intriguing dissociative photodissociation mechanisms, often mediated by a plethora of conical intersections.

The work reported in this Thesis will be of primary relevance to those involved in experimental studies of the thermodynamics, spectroscopy and unimolecular decay dynamics of gas-phase cations. Thermodynamic data are now being obtained with an accuracy close to the resolution of the experiment, *ca.* 0.001 eV or 0.1 kJ mol^{-1}. A definitive value for the adiabatic ionization energy of the CF_3 free radical has been determined from a photoionization study of C_2F_4; this problem which had remained unsolved, despite numerous claims to the contrary, for over 30 years. The thesis shows the enormous potential of *ab initio* techniques, when used in synergy with experiment, to predict molecular properties with high precision; and the value of modelling breakdown diagrams for polyatomic molecular ions using RRKM theory.

This is a superb piece of work by Dr. Harvey and as with many experimental studies in the Twenty-first century in gas-phase Chemical Physics, it would not have been possible without the huge amount of preceding work done by Dr. Andras Bodi and Dr. Melanie Johnson (on the development of the vacuum-ultraviolet beamline at the SLS), Dr. Balint Sztáray, Prof. Thomas Baer and Dr. Patrick Hemberger (on the development of the modelling programmes used throughout this work).

Birmingham, UK, August 2013   Prof. Richard P. Tuckett

With special thanks to John Harvey, George Anderson

Andras Bodi, Michael Parkes, Nicola Rogers and Richard Tuckett

Deity of Science
Acrylic on canvas.
Dimensions 80 × 60 cm

Award winning original art work by the author depicting the iPEPICO endstation and the constant stream of information produced from reactions generated from within.

Contents

1 Introduction and Background Information 1
   1.1 Preamble 1
   1.2 Measuring the Photoelectron Signal 2
      1.2.1 Photoelectron Spectroscopy (PES) 2
      1.2.2 Threshold Photoelectron Spectroscopy (TPES) 3
      1.2.3 Ionization Energies 5
      1.2.4 What Determines the Observed Spectra? 7
      1.2.5 Potential Energy Surfaces 9
   1.3 The Study of Ionic Dissociations 11
      1.3.1 Threshold Photoelectron Photoion Coincidence
           (TPEPICO) 14
      1.3.2 Modelling the Results of TPEPICO Experiments ... 16
      1.3.3 Unimolecular Rate Theories 17
      1.3.4 Kinetic and Competitive Shifts 22
   1.4 Thermochemistry 24
      1.4.1 Thermochemical Values Derived from iPEPICO 24
      1.4.2 Isodesmic Reactions 25
   References 26

2 Experimental 31
   2.1 Preamble 31
   2.2 The Synchrotron Radiation Source 31
   2.3 The Endstation 35
   2.4 Capturing the Electron and Ion Signals 37
   2.5 The Experimental Results 39
   References 39

3 Theory 41
   3.1 Preamble 41
   3.2 Computational Methods 41
      3.2.1 Calculating Molecular Geometries and Potential
           Energy Paths 41
      3.2.2 Composite Methods 42

	3.3	Modelling	46
		3.3.1 Modelling the Breakdown Curves	46
		3.3.2 Modelling the TPES	52
	References	54
4	**Fast Dissociations of Halogenated Methanes: A Thermochemical Network**	57
	4.1	Preamble	57
	4.2	Introduction	57
	4.3	Results and Discussion	61
		4.3.1 Chlorinated Methanes	62
		4.3.2 Fluorinated Methanes	65
		4.3.3 $CBrClF_2$ and $CHClF_2$	66
		4.3.4 Thermochemistry	67
	4.4	Conclusions	77
	References	78
5	**Photodissociation Dynamics of Four Fluorinated Ethenes: Fast, Slow, Statistical and Non-statistical Reactions**	81
	5.1	Preamble	81
	5.2	Introduction	81
	5.3	Results and Discussion	85
		5.3.1 Monofluoroethene	85
		5.3.2 1,1-Difluoroethene	90
		5.3.3 Trifluoroethene	95
		5.3.4 Tetrafluoroethene	97
		5.3.5 Trends and Insights into Bonding	104
		5.3.6 Thermochemistry	104
	5.4	Conclusions	106
	References	107
6	**Threshold Photoelectron Spectra of Four Fluorinated Ethenes from the Ground Electronic State to Higher Electronic States**	111
	6.1	Preamble	111
	6.2	Introduction	111
	6.3	Results and Discussion	113
		6.3.1 Ground Electronic State of the Cations	113
		6.3.2 Electronically Excited Cation States	130
	6.4	Conclusions	140
	References	141

7	**Conclusions and Further Work**	143
	7.1 Conclusions	143
	7.2 Further Work	144
	7.3 Beyond TPEPICO	145
	References	149

Appendix A 151

Appendix B 159

Appendix C 163

Appendix D 169

Chapter 1
Introduction and Background Information

1.1 Preamble

This thesis is divided into seven chapters, the introduction, experimental, theory, three results chapters, conclusion and further work. The results of this thesis are presented in three parts: (i) Chap. 4, the study of *fast* dissociative photoionization reactions of selected halogenated methane cations, (ii) Chap. 5, the study of the *fast* and *slow* dissociative photoionization reactions of larger fluorinated ethene cations and (iii) Chap 6, the study of the fates of the ground and excited electronic states of fluorinated ethene cations. Two themes are prevalent throughout the work, (a) the unimolecular dissociation dynamics of the photoionized molecules and what thermochemical values may be determined from them, and (b) the photoionization of neutral molecules probing the potential energy surfaces and discovering what can be gleaned from investigating the excited states. A summary of the results is presented in the conclusions and further work, in which new directions this work can take are discussed.

This chapter is divided into two main themes. The Sect. 1.2 is concerned with the study of one-photon photoionization. In this section, the techniques used to study these phenomena; photoelectron (PES) and threshold photoelectron spectra (TPES), the process of autoionization, determining ionization energies, the Franck–Condon principle and potential energy surfaces are discussed. The Sect. 1.3 concerns the study of ionic dissociations. In this section, the techniques including threshold photoelectron photoion coincidence measurements (TPEPICO) which is used in this work, how the results are modelled using unimolecular theories, and what kinetic and competitive shifts are. Finally, in Sect. 1.4. the thermochemical values which can be derived from these experiments are discussed.

When a photon is absorbed by an isolated molecule in the gas phase, several reactions can occur. The molecule may become rotationally, vibrationally, translationally and electronically excited. When a molecule is promoted from its ground electronic state to an excited electronic state, the electronic charge gets redistributed. As such, the nuclei can experience a change in Coulomb force and may react to the charge redistribution with enhanced vibration. These transitions are termed *vibronic* transitions and, when part of the ionization process, form the basis

Table 1.1 The different reactions which can occur upon absorption of a photon

	Reaction	Process
(1)	$AB + h\nu_1 \to AB^* \to AB^+ + h\nu_2 + e^-$	Photon emission $h\nu_2 < h\nu_1$
(2)	$AB + h\nu_1 \to AB^+ + e^-$	Ionization
(3)	$AB + h\nu_1 \to AB^* \to AB^+ + e^-$	Autoionization
(4)	$AB + h\nu_1 \to AB^* \to A + B$	Dissociation of the neutral
(5)	$AB + h\nu_1 \to AB^* \to AB^+ + e^- \to A^+ + B + e^-$	Dissociative photoionization
(6)	$AB + h\nu_1 \to AB^* + C \to AC + B$	Reaction with C
(7)	$AB + h\nu_1 \to AB^* \to BA$	Isomerization
(8)	$AB + h\nu_1 \to AB^* + Q \to AB + Q*$	Quenching
(9)	$AB + h\nu_1 \to AB^* \to AB^{\dagger}$	Radiationless transition

Reactions (1), (2), (3) and (5) occur when $h\nu_1$ exceeds the ionization potential

of the research presented in this thesis. Along with the excitation of discrete electronic and vibrational energy levels, accompanying rotational levels may be excited giving the umbrella term for such transitions, *rovibrational* transitions. Translational energy levels are so closely spaced they can be treated as a continuum, and in the situations considered in this thesis, the effect of the photon's momentum on the molecule is imperceptible.

A generic molecule AB can absorb a photon, $h\nu_1$, to become electronically excited, AB*. From that excited state AB* can undergo the following reactions as presented in Table 1.1. The generic molecule AB can also form an ion pair, $A^+ + B^-$, [1] and direct ionization of the neutral to form the ion AB^+ when $h\nu_1$ exceeds the ionization potential, can also occur. Reactions (2), (3) and (5), direct ionization and autoionization and dissociation of the ion AB^+, as well as dissociation of AB^+ formed by direct ionization, are studied in this work using threshold photoelectron spectroscopy (TPES) and threshold photoelectron photoion spectroscopy (TPEPICO). However, the presence of the other reaction pathways (4), (7) and (9) have been observed indirectly.

1.2 Measuring the Photoelectron Signal

1.2.1 Photoelectron Spectroscopy (PES)

Photoelectron spectroscopy (PES) involves the absorption of a photon of electromagnetic radiation by the target molecule, which is of sufficient energy to remove an electron from the bound state in the molecule. The energy limit at which this occurs is the ionization energy (*IE*) and two types of *IE* are generally quoted; when the ion is formed in the zero-point vibrational level ($v^+ = 0$) from the ground state neutral, it has no vibrational energy and the transition is termed the *adiabatic IE (AIE)*. The *AIE* is the lowest *IE* for the transition to a particular ion electronic state. The *vertical* ionization energy (*VIE*) is the most intense member of the vibrational progression in the photoelectron band corresponding to the ground electronic state. That is, it is the most probable transition which may or may not correspond to ionization to the $v^+ = 0$ level, depending on whether the ion geometry is similar to that of the neutral geometry or not.

In conventional PES, ionization is achieved with a fixed energy photon source, where the exact energy is produced by ionization sources such as He I and He II discharge lamps with energies of 21.22 and 40.81 eV. In general the ejection of the valence electrons of a molecule to produce the ion in its different electronic and rovibrational states, is direct and non-resonant [2]. The kinetic energy of the ejected photoelectron is then determined with an electrostatic analyser,

$$hv = IE_i + E_{int} + T_{ion} + T_e \qquad (1.1)$$

IE_i is the adiabatic ionization energy of an electron from a particular orbital i, E_{int} is the internal energy of the ion, T_{ion} is the kinetic energy of the ion and T_e is the kinetic energy of the electron. Most of this excess energy is partitioned into kinetic energy of the electron because conservation of momentum entails that the lighter fragment (in this case the electron) carry it away, so the newly formed larger and heavier cation has very little recoil velocity. The T_{ion} can often be disregarded, and a photoelectron spectrum is generated by measuring the electron current as a function of electron kinetic energy whilst keeping hv fixed. In a molecular orbital picture, electrons are removed from occupied electronic orbitals with increasing energy as orbitals ever closer to the nucleus are probed. All electrons generated contribute to the measured electron signal. Within each photoelectron band corresponding to removal of an electron from each electronic orbital, there is structure originating from the different populations of the vibrational and rotational levels within that electronic state of the ion. With sufficient experimental resolution, this structure can be resolved, unless the rotational envelope is too dominant. Vibrational structure provides the vibrational frequencies of the molecular ion which become active upon ionization. In turn, these frequencies can be used to elucidate the structure of the ion. Not all experimentally observed vibrations are strictly *allowed* under symmetry rules (i.e. only totally symmetric vibrations are allowed), and as will be shown in Chap. 6, these technically *forbidden* vibrational transitions can be used to determine the geometry of the ion. On the other hand if one were able to observe rotational structure it could provide information about the equilibrium structure of the ion, thereby giving not only the geometries but also the symmetries of the electronic states [3]. However this is usually extremely difficult to resolve for all except small molecules.

1.2.2 Threshold Photoelectron Spectroscopy (TPES)

Threshold photoelectron spectroscopy (TPES) follows the general principle outlined above, but only electrons ejected with little to no kinetic energy are detected. The key difference is that, rather than determining the energy of the ejected electron, the detection energy of the electron is fixed and instead the energy of the excited photon is varied. The principal aim of the technique is to probe the electrons associated directly with specific energy levels within the ion, supplying just enough energy to promote the electron to the ionization limit. Of course, electrons with significant kinetic energy (*'hot'* electrons) are also produced in addition to threshold electrons high above the *IE*. As such, suitable discrimination between threshold and hot electron signals is required and the fact that threshold electrons (owing to their negligible kinetic energy) are stationary within the ionization region, whereas hot electrons are mobile, can be exploited. This is further discussed in Chap. 2. Electrostatic analysers which are used in PES are sensitive to Doppler effects which broaden the spectroscopic peaks, limiting the energy resolution. These analysers are not required in TPES experiments, so better resolution can be achieved. However, TPES does require a tuneable photon source such as a synchrotron source, and a dispersive element such as a monochromator (diffraction grating) to select individual wavelengths. The resolution of such an experiment is now limited mainly by the dispersive element. One disadvantage is that access to tuneable radiation sources such as large scale synchrotron sources can be restricted, limiting the autonomy of the experimentalist. In addition, diffraction gratings disperse the light over a number of orders which can contaminate the signal, but this can be dealt with easily using a variety of filters (discussed in Chap. 2).

Ions and electrons can be formed indirectly via a multi-step pathway e.g., autoionization, and can be detected with resonant techniques such as threshold photoelectron photoion coincidence (TPEPICO) and TPES. The general process is described by the following scheme

$$hv + AB \rightarrow AB^* \rightarrow AB^+ + e^- \qquad (1.2)$$

The neutral molecule absorbs a photon and is excited into a high-lying neutral electronic state, AB*, usually a Rydberg state, i.e. an ion core with the excited electron in an orbital with a high principlal quantum number. Therefore the electron in the Rydberg orbital spends the majority of its time at large distances away from the core, seeing it as a point charge [4]. A dense series of these Rydberg states or a quasi-continuum converges upon each ion states. The excited neutral state can decay via *predissociation* by crossing over onto a neutral dissociative state, producing no charged species and only neutral fragments which are difficult to detect in typical coincidence and threshold photoelectron measurements. It can also decay to a lower lying neutral electronic state whilst emitting a photon (fluorescence) which is again undetectable using coincidence and photoelectron techniques. Finally autoionization can occur via a radiationless transition (see Fig. 1.1.) [5], the Rydberg state can couple to a lower energy ion state, producing a vibrationally or rotationally excited ion and the corresponding ejected threshold electron [6]. Therefore, additional spectral features can be identified in the TPES in between electronic bands, i.e. within Franck–Condon gaps, where there is no direct overlap between the states involved in the transition. These states within the

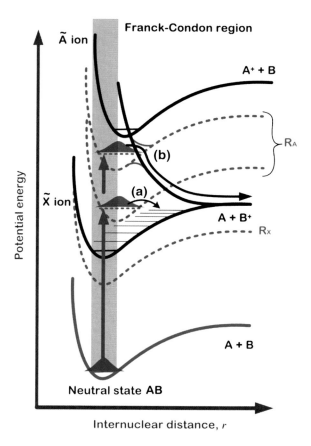

Fig. 1.1 Potential energy curves showing **a** autoionization. Neutral Rydberg orbitals converge upon the excited $\tilde{A}$ state of the ion. The neutral molecule is excited to the Rydberg orbital R_A which can cross over to the ground electronic ion state, $\tilde{X}$, creating vibrationally excited $\tilde{X}$ state ions and an ejected electron with low kinetic energy. **b** Predissociation, where a dissociative potential crosses an excited Rydberg orbital to produce ion and neutral fragments

gap regions are not accessible using non-resonant ionization techniques, unless by chance, the incident light has the same energy as the energy required for the transition to a Rydberg state.

1.2.3 Ionization Energies

In instances where there are no overlapping features in the photoelectron spectra, the ionization energy to a particular orbital can be found which corresponds to each individual photoelectron band (Fig. 1.2). However, in instances of spectral congestion, identification is not straightforward, and ionization energies can be calculated instead. Two theoretical approaches can be taken, treating the electron in terms of its specific wave function (ab initio Hartree–Fock method) or considering the electron density (the semi-empirical density functional theory).

The numerical solution to the Schrödinger equation gives a description of atomic orbitals and can be found using the ab initio self-consistent field (SCF) method developed by Hartree, and improved upon by Fock and Slater (HF-SCF) [7].

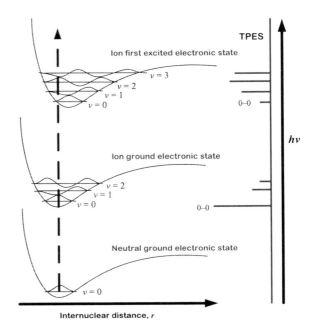

Fig. 1.2 Schematic of neutral and ion Morse potential energy curves together with a stick TPES illustrating the Franck–Condon principle in action. The dashed line represents the excitation photon. Maximum overlap occurs between the vertical transition between the $v = 0$ wavefunction of the neutral ground state and $v = 0$ wavefunction of the ion ground state, the origin peak 0–0. The first excited state has a longer equilibrium bond length and as such the overlap integrals are different resulting in a maximum which is not at the 0–0 but at the 2–0 peak

Ignoring electron–electron repulsions, the overall wave function for n electrons can be expressed as the product of n single-electron wave functions, and is dependent upon the nuclear locations [8]. The overall energy is then given by the sum of the single-electron energies. Within this framework, Koopmans' theorem is derived; the energy required to remove an electron from its orbital is equivalent to the negative of the energy of that orbital [9]. It is assumed that the remaining electrons do not rearrange in response to the electron removal. Spinorbitals, which are products of an orbital wavefunction and a spin function, are introduced to account for electron spin ensuring the wave function obeys the Pauli principle [7]. However, the effects of electron–electron repulsions must be considered. In HF-SCF, these repulsions are treated in an average way where each electron is regarded as moving in the average field of the other $n - 1$ electrons. In the HF equations for the individual spinorbitals which give the wave function, there is a Coulomb operator that accounts for the repulsions. This operator often over estimates the contribution made by the repulsions to the spinorbital energy because the correlated motion of the electrons is not accommodated. An exchange operator, which accounts for the effects of spin correlation is also included in the HF equations. Finding the numerical solution to the HF equations for molecules is computationally complex so the molecular orbitals are expanded as a linear combination of atomic orbitals (Hartree–Fock–Roothaan equations) [10]. A known set of basis functions are used to expand the spinorbitals as precise numerical solutions of HF equations for atoms are not useful in this context; the molecular spinorbital is equal to the sum of the basis functions multiplied by an expansion coefficient for that orbital [7]. Now, an initial guess of the coefficients are made and the HF equations are solved generating a new set of coefficients and orbitals which are used to solve the HF equations, and so on, repeating until there is no change in the solutions between iterations. The calculation is said to be converged and the solutions are self-consistent [8]. Assuming that no change in electron correlation (which is neglected in HF methods) occurs upon ionization can lead to an underestimation of the ionization energies. Wave function theories (WFT) e.g. coupled cluster methods, are post HF methods which are computationally more expensive than HF but seek to accommodate the effects of electron correlation, thereby facilitating calculations of open shell molecules and higher excited electronic states [11].

An alternative to HF methods is density functional theory (DFT) which is based on the Hohenberg–Kohn theorem and states that the ground-state energy and properties are determined by the electron density [7]. DFT primarily deals with electron density rather than the electronic wave function as with HF theory. In DFT, the energy of the electron system can be expressed in terms of electron probability density to give Kohn–Sham (KS) orbitals [12]. The KS orbitals can be computed numerically or expressed in terms of a set of basis functions where the ground state electron density is given as the sum of the occupied KS orbitals squared. The electronic energy is a functional of the electron density, and incorporates the kinetic energy of the electron, the electron–nucleus attraction and the Coulomb interaction between the total charge distributions (summed over all KS orbitals) plus the exchange–correlation energy of the system [8]. The term for the exchange–correlation energy includes all non-classical electron–electron interactions and approximations are needed for its derivation. Unlike the HF method, in which it is neglected, DFT uses a semi-empirical approach to electron correlation. If one could derive the exact Kohn–Sham orbital, then the negative of the highest occupied molecular orbital energy would correspond to the vertical ionization energy [12, 13]. The success of DFT relies upon the type of exchange–correlation functional chosen, but the reduction in computational cost for larger systems compared with traditional wave function methods makes it very attractive to use. However, neither method are as reliable for calculating excited electronic states as for ground electronic states [7].

1.2.4 What Determines the Observed Spectra?

Vibronic transitions as observed in TPES experiments are a combination of electronic, vibrational and rotational transitions which arise when an electronic transition occurs from the neutral molecule to the ionized molecule. The probability of an observable transition occurring can be given by the Franck–Condon principle [14]. This principle follows on from the Born–Oppenheimer approximation; the masses of the nuclei are so much larger than the mass of the electron involved in the transition, the electron moves between the orbitals involved so swiftly that the nuclear locations are held to be virtually identical, before, during, and after the transition. Put another way, the time period for electronic promotion is considerably less than the time it takes for the nuclei to vibrate; the inter-nuclear separation in the upper state is assumed to be the same as that of the lower state [8]. As a result of this assumption, the molecule makes a vertical transition (see Fig. 1.2). The Franck–Condon factor determines the vibrational contribution to the transition probability. If the nuclei are not extensively displaced from the equilibrium position, excluding the nuclear contribution to the electronic dipole moment μ_n (which has become zero) the transition probability is as follows [15];

$$M = \int \psi'_e(r, R_e) \cdot \mu_e \cdot \psi''_e(r, R_e) dr \quad \int \psi'_v(R) \cdot \psi''_e(R) dR \quad (1.3)$$

M is the transition moment, μ_e is the electronic part of the dipole moment operator, R_e is the equilibrium coordinates, r and R are the electronic and nuclear coordinates respectively, and ψ_e is the electronic wave function, with prime and double prime denoting upper and lower states respectively. In the context of this work, the upper refers to the ion state and the lower, the neutral state. The first integral supplies the basis for the electronic selection rules and determines the overall intensity of the transition. The second integral determines the intensity of the individual vibrational transitions within an electronic transition. This second integral is an overlap integral for the vibrational wave functions of the upper and lower electronic states. The square of this second integral gives the Franck–Condon factor (*FCF*) [15];

$$FCF = \left(\int \psi'_v \psi''_v \, dR \right)^2 \quad (1.4)$$

Franck–Condon factors can be used to determine the probability of transitions to different vibrational levels while taking into account the change of the geometry between lower and upper states [8]. Transitions with a high probability involve a transition from lower vibrational states to upper vibrational states which have similar vibrational wave functions, thus providing maximum overlap between the two vibrational states, and manifest themselves as intense peaks in the TPES. It follows that vibrational transitions with wave functions that have less overlap are seen as weaker peaks within a band of the TPES, and if the FC is zero then no peak is seen, see Fig. 1.2 [3].

Whilst for diatomic molecules, all vibrational wavefunctions are totally symmetric, for polyatomics this is not the case as not all normal modes have totally symmetric normal coordinates.[1] However, modes with totally symmetric normal coordinates usually provide the majority of the vibrational structure seen in

---

[1] A normal mode is a vibrational mode where all the nuclei harmonically vibrate with the same frequency, preserving the centre of mass, moving in-phase but generally with different amplitudes along the normal coordinates. The normal coordinate system is an alternative set of coordinates apart from Cartesian coordinates for each atom in a system, which removes cross-terms (coupling) in either the kinetic and potential energy operators.

electronic spectra (including TPES). Totally symmetric modes are prevalent throughout electronic spectra because the vibrational wave function for a normal vibrational mode with a totally symmetric normal coordinate is totally symmetric for all vibrational quantum numbers v. This gives a *FCF* with an integrand which is totally symmetric [8]. Modes which do not have a totally symmetric normal coordinate have vibrational wave functions which oscillate between being totally symmetric and non-totally symmetric, as the levels change from even to odd [15]. Therefore, the *FCF* has a non-totally symmetric integrand, when either the lower or upper state vibrational wave function is non-totally symmetric. It then follows that if a transition occurs from the lower totally symmetric vibrational state to the second vibrational level, $v^+ = 2$, of an upper state which is non-totally symmetric, the integrand becomes totally symmetric as the wave function at $v^+ = 2$ *is* totally symmetric [8]. Examples of this are given in Chap. 6. Changes in the geometry upon ionization along the normal coordinate excite the vibration. For example, it can be seen in Chap. 6 that upon ionization, the C=C bond length in the fluorinated ethenes increases, and the major most intense vibrational progression is the C=C stretch. If the neutral and ion molecules have significantly different geometries then the TPES will not exhibit a sharp $v = 0$ to $v^+ = 0$ transition between totally symmetric vibronic states, but a broad Franck–Condon envelope. Finally, if the geometry of the upper state changes significantly from the lower state, then transitions between totally symmetric and non-totally symmetric vibrational states can become viable because of vibronic coupling, occurring with a larger FCF and hence degree of probability. This was found to be the case for 1, 1-$C_2H_2F_2$, see Chap. 6.

1.2.5 Potential Energy Surfaces

The neutral and ion molecules can be thought of in terms of a multi-dimensional *potential energy surface*, for polyatomic molecules this is a landscape formed by the electronic energy as a function of $3N - 5$ or $3N - 6$ (linear and non-linear molecules respectively) internal nuclear coordinates. The Born–Oppenheimer approximation already mentioned in Sect. 1.2.4, states that the nuclear and electronic components of the wave function can be separated [16]. The result is, the molecule will be found in its original state when the perturbation ceases [17]. The full wave function can be expressed as a product of the electronic and nuclear wave functions. For the approximation to hold, the timescale of the perturbation must be longer than the timescale for adjustment and the same electronic state is maintained throughout [18]. For ease of visualization, *potential energy curves*, slices through the surface charting how the energies change with respect to one coordinate as shown in Fig. 1.3, are often considered. At each point on this surface the nuclei are frozen, and the lowest eigenvalue and stationary electron wavefunction for that configuration are determined. Such paths are termed *adiabatic*, see Fig. 1.3a.

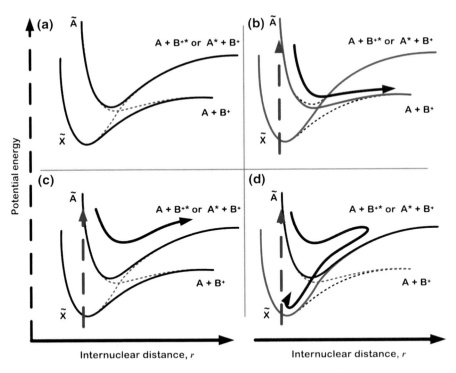

Fig. 1.3 a Adiabatic (*solid line*) and diabatic (*dashed lines*) potential energy curves for the dissociation of a generic system, AB$^+$ **b** Curve crossing following the non-adiabatic path giving ground state products. **c** No crossing between the adiabatic paths occurs, products are produced in their excited state. **d** Some of the wave packet may travel onto the lower state through the crossing, into the potential well of the ground state, $\tilde{X}$, which can then form other product ions other than A$^+$ or B$^+$. The large *dashed blue arrow* represents the Franck–Condon excitation to the upper state of the ion $\tilde{A}$ from the lower energy ground electronic state of the neutral (not shown)

As the shape of the potential energy surface depends on the *electronic* energy, different electronic states give rise to different potential energy surfaces for the nuclei, as shown in Fig. 1.3. If the system described above is not allowed to re-adjust to the effects of perturbation, i.e. the electrons cannot rearrange quickly enough in response to fast nuclear vibration, the nuclear and electronic motions are not fully separated. This is termed vibronic coupling and manifests as a conical intersection (a funnel) between the two electronic states involved in the transition, Fig. 1.4 [18]. It follows that if the energy difference between the two states is sufficiently large then coupling does not occur and the Born–Oppenheimer approximation holds. However, as the energy separation decreases, the probability of coupling increases. Conical intersections allow the electronic state of the system to change without a change in the kinetic energy of the nuclei [8]. It has been found that bending and torsional modes of polyatomic molecules can facilitate

1.2 Measuring the Photoelectron Signal

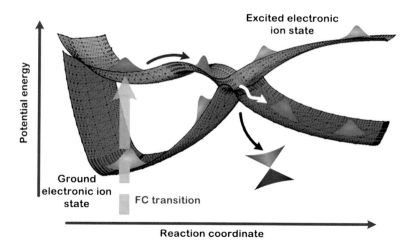

Fig. 1.4 Schematic of 3-D potential energy surfaces of the ground electronic state and first excited electronic state. The wavepacket is transposed onto the ion manifold via a Franck–Condon transition from the ground electronic state of the neutral At the conical intersection the reaction can follow the lower path to form electronic ground state products (*red*), reflect into the ground state well or stay on the upper state (*grey*)

electronically diabatic transitions [8, 19–22]. Diabatic behaviour is now considered ubiquitous for many systems, [23] and can also be invoked to explain the apparent statistical behaviour of systems that initially were perceived as non-statistical [21]. As Fig. 1.4 shows, the system once formed in the excited state can stay on the upper state or cross back down to the ground state. From here, it can either form ground state products or flow back down to the ground state potential well and go on to form other products. These two aspects are investigated in Chaps. 5 and 6.

1.3 The Study of Ionic Dissociations

Understanding how cations, as discussed above, decompose into their respective daughter ions and neutrals following vacuum ultraviolet (VUV) photoexcitation; their so-called unimolecular dissociation dynamics, is of fundamental interest. Such studies try to answer the following questions; what are the branching ratios between the different decay pathways, what are the absolute rates of decay and how is the energy partitioned in the product channel? Several experiments have been developed over the years to study such unimolecular dissociations.

Photoionization Mass Spectrometry (PIMS) involves measuring the mass analysed ion signal as a function of photon energy [24, 25]. The generated spectrum is a Photoionization Efficiency (PIE) curve. The main piece of information derived from PIMS experiments is the appearance energy of an ion at the temperature of the experiment, which is usually 298 K ($AE_{298\,K}$) [26]. In order to determine the $AE_{298\,K}$, a straight line is fitted to the lower energy part of the curve which is linear and it is then extrapolated down to zero [27]. The $AE_{298\,K}$ can then be used to deduce unknown thermochemical parameters [26], e.g. the enthalpies of formation for neutrals, radicals and ions, and bond dissociation energies. For the following generic reaction;

$$AB + h\nu \rightarrow AB^+ \rightarrow A^+ + B + e^- \quad (1a)$$

the $AE_{298\,K}$ is approximately given as;

$$AE_{298K}(A^+) = \Delta_f H^\theta_{298K}(B) + \Delta_f H^\theta_{298K}(A^+) - \Delta_f H^\theta_{298K}(AB) + E^* \quad (1.5)$$

where $\Delta_f H^\theta_{298\,K}$ is the enthalpy of formation at 298 K, and E^* is the total excess energy available to the system after dissociation, composed of the kinetic energy of the ejected electron and the kinetic and internal energy of the two fragments A^+ and B.

When the onset is sharp, the appearance energy is relatively straightforward to determine. However, if the onset is broadened by experimental factors such as photon intensity, sample pressure, competing reactions coupled with weak Franck–Condon factors for the ionization process, or a broad ion internal energy distribution, then determining the precise value of $AE_{298\,K}$ is more troublesome [25, 27]. Another limitation of this method is that no information about the internal energy of the parent ion AB^+ can be given directly because the energy of the photoelectron is unknown. This uncertainty arises because we do not necessarily know how much of the excess energy has been partitioned into the kinetic energy of the electron. Kinetic shifts (see later) are not easily observed and so no information about reverse barriers or slowly dissociating ions is obtained [24].

An improvement upon PIMS experiments is Photoelectron Photoion Coincidence spectroscopy (PEPICO) in which the molecule is directly ionized using a fixed energy light source, it ejects an electron and the mass of the corresponding (coincident) ion is measured by time-of-flight mass spectrometry [28]. Unlike PIMS where only the ions are detected, both of the charged particles are subsequently measured in correlation to each other as a function of the electron kinetic energy. Plotting the fractional parent and daughter ion abundances against the electron kinetic energy generates the breakdown diagram [29]. In this technique, all electrons are correlated to their associated ions and are detected. Ions resulting from ionization of the initial neutral molecule, termed parent ions, and positively charged ion fragments (daughter ions) resulting from subsequent dissociation of that parent ion are detected. The excitation source is usually of a fixed energy e.g. the He I discharge lamp at 21.22 eV, and the electrons are detected with an electron analyser. The energy of the ions is then examined by varying the energy of the electron that is collected [30]. The fixed energy excitation sources are high in energy, producing ions in a large distribution of internal energy states. The fraction of ions produced in the energy band pass compared to all ions produced is small, leading to a large false coincidence signal,

1.3 The Study of Ionic Dissociations

i.e. coincidences between ions and electrons which do not originate from the same event [31]. In addition, only small quantities of electrons are actually ejected towards the electron monochromator, and so collection efficiencies of hemispherical electron analysers are very low. Consequently, the overall collection efficiency for the experiment is low, i.e. less than one in every 1000 ions are detected in coincidence with their electron [29, 32].

By detecting the energy of the electrons at a fixed photon energy, we can determine the internal energy of the ions produced from that same event from Eq. 1.1 [32]. In order to detect both the ions and electrons, they need to be extracted towards their respective detectors. This is achieved by applying a small electric field, and it is this field which greatly affects the electron energy resolution. Good electron resolution requires low extraction fields, but improved ion mass resolution is obtained with higher extraction fields [29].

PEPICO is a direct precursor to Threshold Photoelectron Photoion Coincidence spectroscopy (TPEPICO) which is the technique used within this work. Threshold in the above acronym refers to the detection of only electrons with virtually zero kinetic energy imparted into them, and by consequence, their corresponding internal energy selected ions [29]. Only threshold electrons and coincident ions are detected, as opposed to scanning the electron energy as with PEPICO, and offering internal energy selection unlike with PIMS. Tuneable energy sources are used to produce threshold electrons just at the ionization limit by a combination of *direct* (to the ionization limit) and *indirect* ionization (via autoionization from neutral Rydberg states) processes. Time-of-flight (TOF) mass spectrometry is a widely used method for ion mass detection. The ejected electron has a significantly smaller mass than the ion and reaches its detector first to start a clock, which is then stopped by its corresponding ion reaching its detector some time period (typically in the order of tens of μs) later. It follows that heavier ions will take a longer time to reach the detector than smaller ion fragments. With this technique, the 0 K appearance energy can be directly determined either by inspection of the experimental ion yields or by modelling them [29]. One advantage threshold electron detection offers is better electron energy resolution (less than 1 meV) because the need for dispersive electron analysers as used in PEPICO is removed, as only threshold electrons are selected. Greater sensitivity is also afforded because practically all threshold electrons reach the detector. Therefore, the signal to noise ratio is increased and false coincidences are reduced. Traditional PEPICO is restricted to exploring dissociations from electronic states of the ion which are accessible with a vertical transition from the ground electronic state of the neutral, via direct ionization. In other words only those states which are Franck–Condon accessible can be investigated. Molecules which dissociate within Franck–Condon gaps are not accessible by direct ionization, and so their onset energies cannot be measured with traditional PEPICO methods. TPEPICO on the other hand can measure these onset energies (see Chaps. 4, 5), where the ejection of a threshold electron often corresponds to formation of a vibrationally excited ion [31]. Comparisons between the threshold photoelectron spectrum (TPES) which is also sensitive to autoionization processes, and a photoelectron spectrum (PES) obtained using a non-resonant discharge lamp e.g. He I PES, reveal the extent of autoionization [24, 29].

The ions can be extracted from the ionization region either by the use of a pulsed ionization source with continuous ion extraction e.g. a VUV laser [33, 34], pulsed ion extraction [35, 36] or continuous ion extraction [31, 37]. Synchrotron radiation as used throughout this work is quasi-continuous and only the latter two methods can be used. Another variation on PEPICO is Pulsed Field Ionization (PFI) PEPICO, such as that established by Ng and co-workers at the Advanced Light Source (ALS) in Berkeley, California [35]. PFI PEPICO in which high-n Rydberg states of the molecules are field ionized, allows very accurate measurements of 0 K appearance energies to be made. The multi-bunch mode of the ALS consists of 272 bunches of electrons per orbit of the storage ring, each lasting 50 ps and separated by a gap of 2 ns, has a dark gap of 112−140 ns at the end of the ring period [35, 36]. Molecules are field ionized during this dark gap, and prompt electrons are distinguished from those originating from field ionization. A very high resolution photon monochromator is needed to excite the very narrow band of Rydberg states just below the ionization limit [29, 35]. Experimental parameters such as the height of the Stark pulse and delay with the start of the dark gap are adjusted so that no hot electrons are observed, providing discrimination of the pure PFI electron signal [35, 38]. The improved electron resolution afforded by this technique of 0.1 meV was achieved by removing the need for a high Stark pulse [39]. However, the diabatic nature of the field ionization of the high-n Rydberg states limits the resolution of the experiment [33, 40]. Also, the consequence of using these very low electric fields mean that the ions are not efficiently extracted from the ionization region. This fact combined with the absence of field ionization from *long-lived* Rydberg states means the signal to noise is reduced and ion TOF distributions are not as well defined as with stronger fields. Consequently slow dissociations cannot be analyzed [38].

Further experiments involving coincidences between; ions/electrons and emitted photons (PIFCO) and (PEFCO) [41, 42], ions and ions resulting from dissociation of a doubly charged parent ion (PIPICO) [43] and between electrons and both fragment ions (PEPIPICO) [44, 45] and more recently, multi-electron coincidence (PEPEPICO) [46] experiments, have also been developed to explore the wide range of molecule-photon interactions.

1.3.1 Threshold Photoelectron Photoion Coincidence (TPEPICO)

A more detailed examination of TPEPICO spectroscopy will now be given. Threshold coincidence experiments involve the measurement of internal energy selected ions and their corresponding close to zero-energy (threshold) ejected electron. The measured ions include the initially ionized molecule (parent ion,

Reaction 1b) and subsequent fragmentations into secondary and tertiary ions (daughter ions, Reaction 1c and 1d).

$$ABC + h\nu \rightarrow ABC^+ + e^- \tag{1b}$$

$$ABC + h\nu \rightarrow ABC^+ \rightarrow AB^+ + C + e^- \tag{1c}$$

$$ABC + h\nu \rightarrow ABC^+ \rightarrow AB^+ + C \rightarrow A^+ + B + C + e^- \tag{1d}$$

The internal energy of the ion ($E_{int}(M^+)$) is given as

$$E_{int}(M^+) = E_{int}(M) + h\nu - IE_{ad} \tag{1.6}$$

where $E_{int}(M)$ is the internal energy of the neutral, $h\nu$ is the photon energy and IE_{ad} is the adiabatic ionization energy. This holds if the internal energy distribution of the neutral is faithfully transposed upon the ion manifold following ionization, i.e. all neutral molecules have equal threshold ionization cross-sections. In the absence of tunnelling, E_0 is the onset energy at zero kelvin, in other words the threshold energy for the dissociative photoionization reaction or barrier height [47–49]. Unimolecular dissociation reactions of internal energy selected parent ions are studied as a function of photon energy, yielding 0 K daughter ion appearance energies, E_0 [37]. The breakdown curves for the parent and daughter ions would be step functions for fast dissociations (the parent ion yield decreases from 100 to 0 %, and daughter ions increases from 0 to 100 % at one single energy) if only one internal energy mode in the neutral molecule were populated. However, in most experiments the neutral sample has an initial thermal (Boltzmann) energy distribution where the majority of the energy is partitioned into rotational modes, giving a broader breakdown diagram. The curves usually change monotonically resulting in smooth rises and decreases in ion signals. Under such circumstances, it can be assumed that the threshold cross sections and collection efficiencies are constant over the threshold energy range and the derivative of the breakdown diagram yields the thermal energy distribution [50]. However, if there are peaks seen in the breakdown curves, then this assumption is not valid. These peaks may correspond to photoionization to rovibrational and/or electronically excited ion states producing ions which have less internal energy than photoionization to the ground electronic state at the same photon energy [50].

For fast dissociations, E_0 is found where the parent ion signal reaches zero. For slower dissociations however, this value may be obscured by a *kinetic shift* and may appear some several hundred meV below the actual disappearance of the parent ion signal.

1.3.2 Modelling the Results of TPEPICO Experiments

For an ionized molecule under collision-free conditions, the density of electronic states is generally higher than that for neutral molecules [51]. Consequently, after photoionization and aided by conical intersections, the excess energy is rapidly and randomly redistributed amongst the vibrational degrees of freedom of the *ground* electronic state of the molecular ion [51]. This is because the density of states of the ion ground electronic state typically exceeds that of any higher lying excited states [31]. Statistical redistribution is usually much faster (on a timescale of 10^{-10} s) [52] than other possible parallel processes such as radiative processes (infrared radiative decay) [29], which occur on a longer timescale of ms. Thus, the long lived intermediate state has the opportunity to sample all the available phase space of the dissociating ion. As a result of this energy randomization, subsequent dissociations of the ion can be studied using statistical rate theories such as Rice–Ramsperger–Kassel–Marcus (RRKM) theory, because the ion has now 'forgotten' how the energy was partitioned in its initial preparation.

The contrasting scenario is one with reduced vibrational couplings and/or reduced surface crossings between the excited and ground states of the ion. In this instance, the state does not sample the whole of the energetically allowed phase space of the dissociating species, and redistribution into all available vibrational modes does not occur. Dissociation can occur from such trapped states and are termed '*isolated states*'. Usually this can happen if the dissociation rate is faster than the internal redistribution of energy [31]. Fragmentation pathways from isolated states are often specific to particular *excited* electronic states of the ion, so their ion yields tend to follow the TPES. An example of non-statistical decay is impulsive atom loss. The neutral molecule is excited to a repulsive ion state and internal redistribution of energy cannot take place before it directly dissociates into daughter fragments. Formation of the ground electronic state of the parent ion is essentially by-passed [53, 54], and dissociation is a non-statistical process. In this instance, fragment ions are accompanied by large kinetic energy releases (KER), in an explosive decay [31, 51].

However, not all isolated states are unbound. If the molecule is excited to a long-lived bound higher electronic state, and that higher state is prevented by whatever reason from converting back down to the ground state, then this upper state is also isolated. Dissociation occurs from this excited state and the ion yield follows the TPES. Due to the long-lived nature of the excited electronic state, dissociation can be treated as a statistical process within its own subspace, where all but the ground state pathways are accessible [51]. Evidence of such decay from long-lived but isolate states is given in Chap. 5 for the loss of an F-atom from $C_2F_4^+$, where the breakdown curves for this reaction and the sequential loss of CF_2 from $C_2F_3^+$ is successfully modelled using statistical rigid-activated-complex RRKM theory. In Chap. 6 the division of the reaction flux between statistical and non-statistical pathways in $C_2H_3F^+$ is explored.

1.3.3 Unimolecular Rate Theories

The breakdown diagram generated from the TPEPICO data and so the rates at which unimolecular dissociation occurs can be modelled using the statistical Rice–Ramsperger–Kassel–Marcus (RRKM) [55] theory to extract the 0 K onset energy, E_0 [52, 56]. This theory can also be used to successfully model consecutive and parallel reactions [37, 52, 57].

The basic assumptions of RRKM theory are; (i) the translational, rotational, vibrational and electronic motions within the reacting system are separated. (ii) Nuclear motion is adequately described using classical mechanics, and where appropriate corrections made based on quantum mechanics. (iii) The reaction crosses through the transition state between reactants and products only once, and a return journey does not occur. (iv) A state of quasi-equilibrium exists where the density in phase space of the excited molecules is uniform prior to reaction, and is independent of the mode of preparation. The result of applying the aforementioned assumptions is that the rates can be determined in a straightforward manner, removing the need to sample the initial conditions and solve equations of motion for a large ensemble of trajectories (which are both difficult and time consuming) [52].

Reactants in equilibrium over a narrow energy range possess a microcanonical distribution. Within this narrow energy range, all microstates of the reactant are equally probable and so all the ways of partitioning the available energy between the internal (bound) $3n - 7$ degrees of freedom of the transition state and the reaction coordinate translational motion are equally probable. This is because each reaction path which follows through to products has originated in the reactant region. Each quantum state at the transition state can be correlated to such a state in the state of reactants and all quantum states of the transition state are equally probable. From this assumption, it follows that the rate of surmounting the reaction barrier through the transition state configuration will be fast when significant amounts of energy are partitioned into translational degrees of freedom (along the reaction coordinate). The rate will be slower if the majority of the energy is partitioned into other degrees of freedom and a statistical treatment of the dynamical problem may be given [17]. Generally the assumption is made that the various excited electronic states, which are initially populated in the ionization process, rapidly decay to the ground ionic state via internal conversion where all of the electronic energy is converted into vibrational energy of the ground electronic state. However, the electronic and vibrational energy can be interconverted between the many different electronic states that lie below the dissociation limit. Though it has been found that contributions to the density of states from these other excited states is often negligible [58].

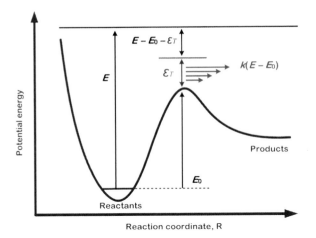

Fig. 1.5 Reaction coordinate for a dissociation with a reverse barrier. The transition state is found at the top of the barrier. E is the total ion energy and E_0 is the activation energy. The ε_T is the translational energy in the reaction coordinate. $E - E_0 - \varepsilon_T$ is the energy avaliable which is statistically distributed. The dashed line is the zero point energy

1.3.3.1 Rice–Ramsperger–Kassel–Marcus (RRKM) Theory

Figure 1.5 shows the reaction path coordinates for moving the reactants through to products via a barrier on which is located a saddle point, or in other words a transition state. The saddle point makes an excellent choice for a transition state because at this least stable *configuration of no return* (following on from assumption (iii) it takes only the smallest change in the initial conditions of the reaction trajectory to take it through to products [57]. The top of the barrier is the *bottleneck* (true transition state) in the phase space between the reactants and products, where the flux of trajectories across it is minimal. The rate of flux through this transition state is greater than or equal to the true reaction rate.

The total energy at the transition state boundary, E, is given by,

$$E = E_0 + \varepsilon_T + \varepsilon_I \tag{1.7}$$

where E_0 is the purely electronic energy barrier minus the zero point energy (zpe) of the reactants plus the zpe of the internal modes at the transition state, ε_T is the 1 dimensional translational energy along the reaction coordinate, R, and ε_I is the internal energy of the transition state. The rate of passage across a barrier for a 1-D translational state, dr^+ is,

$$dr^+ = v^+ dN^+ = v^+ \frac{dp^+}{h} = \frac{d\varepsilon_T}{h} \tag{1.8}$$

where dN^+ is the number of systems per unit length along R, denoted by $dN^+ = dp^+/h$. dp^+ is the linear momentum along R and v^+ is the linear velocity along R [17].

The rate of passage across a barrier for a 1 D translational state is equal to the rate of passage through the *point of no return* (i.e. the saddle point), $d\varepsilon_T/h$ along q, irrespective of the velocity and therefore ε_T itself. So, when the total energy is in

1.3 The Study of Ionic Dissociations

the range E to $E + dE$ and the internal degrees of freedom are in a given state then the rate of crossing the barrier is universal ($1/h$ per unit translation energy), irrespective of the particular details of the reactant. A given total energy, E, may be partitioned many ways between the internal energy of the transition state, ε_I and the translational energy along R, ε_T. An important assumption is made, that at equilibrium partitioning the energy into each state is equally probable. From this assumption, when the total energy is in the range E to $E + dE$, the rate of passage over the barrier, n, is the sum over internal states of all the rates of crossing,

$$n = \sum_{\substack{internal \\ states}} dr^+ = \sum_{\substack{internal \\ states}} \frac{d\varepsilon_T}{h} = \sum_{\substack{internal \\ states}} \frac{dE}{h} = \frac{dE}{h} \sum_{\substack{internal \\ states}} 1$$

$$\equiv \frac{dE}{h} N^{\ddagger}(E - E_0) \tag{1.9}$$

The above equation refers to a particular energy state of the reactant, n. The sum encompasses all internal states whose internal energy is within the allowed range from 0 to $-E_0$ and for a given ε_I, $d\varepsilon_T = dE$. It follows that the rate of passage for all reactants with a total energy in the range E to $E + dE$ is given by the number of internal states $N^{\ddagger}(E - E_0)$ of the molecule at the transition state whose internal energy ε_I is in the allowed range $0 \leq \varepsilon_I \leq E - E_0$. Dividing the rate of crossing by the concentration of the reactants delivers the reaction rate constant,

$$k(E) = \frac{\sigma N^{\ddagger}(E - E_0)}{h\rho(E)} \tag{1.10}$$

where h is Planck's constant, σ is the reaction path degeneracy, $N^{\ddagger}(E - E_0)$ is the transition state sum of states from 0 to $E - E_0$, $h\rho(E)$ is the parent ion density of states at energy (E). The superscript ‡ indicates that one degree of freedom is absent, so $N^{\ddagger}(E - E_0)$ only represents the bound internal states of the transition state. This means that only real vibrational frequencies are included (for non-linear molecules, $3n - 7$ vibrational modes) and imaginary frequencies (the unbound motion that carries the molecule through to its product state) are excluded [31, 56, 57]. When $E = E_0$, Eq. 1.11 is produced giving the threshold rate constant, the lowest value possible to $k(E)$.

$$k(E_0) = \frac{1}{h\rho(E_0)} \tag{1.11}$$

Ignoring the rotational energy of the ion, the sum and density of states refer just to the vibrational degrees of freedom. A molecule will have a total number of s vibrational degrees of freedom (if n is the number of atoms, then $3n - 6$ degrees of freedom for linear transition states, and $3n - 7$ for non-linear transition states) and an internal energy E (measured from the molecule's zero point energy), then the sum of states is all the possible ways in which to distribute the energy amongst s number of oscillators where the total energy is less than or equal to E. For example, all oscillators are in their lowest energy state (the ground state) in one vibrational configuration, another configuration could be where all but one (or two, three etc. up to s number of oscillators) are in their ground state and the remainder are in some excited vibrational state. The number of those states with rovibrational energy less than or equal to E can be added up by the Beyer–Swinehart direct count algorithm [59] to give the number of states, $N^{\ddagger}(E - E_0)$, which can be viewed as a measure of how loose the transition state is. A loose transition state corresponds to a large number of states giving rise to fast rates. As the energy increases, so too does the number of ways in which to distribute the energy. The difference between RRK and RRKM theory can now be highlighted. In RRKM theory the different vibrational frequencies and hence different energy content is recognized for each oscillator and therefore must be treated quantum mechanically. The density of states can be considered as the derivative of the sum of states. The density of states is the number of vibrational configurations with an energy content between E and $E + \delta E$.

The direct count method used to determine both the number and density of states is only suitable for small ions and for energies close to threshold. In ionic systems with many more atoms and at high energies, direct count methods become greatly complex and time consuming so that alternative methods must be sought. For higher energies the Whitten–Rabinovitch [60] approach is more suitable. However, it is less suitable for larger molecules because the number of vibrational quanta excited at energies just above threshold is small. Therefore large ions with many oscillators are much closer to the quantum limit than small ions. The Whitten–Rabinovitch method incorporates a scaled zero point energy into the classic mechanical equation [60]. Another method is the steepest descent method and is based on the inversion of the partition function. The density of the states is obtained by solving the inverse Laplace transform, but though mathematically complex, can be computed using just 10 lines of computer program code [57, 61]. Modelling the breakdown diagrams obtained from the TPEPICO experiments to obtain the onset E_0 provides an excellent opportunity to test the validity of RRKM theory. It confirms the proposed assumptions, that the neutral thermal distribution is faithfully transposed onto the ionic manifold, that statistical energy redistribution is complete and all oscillators participate equally, and as such, further refinements such as incorporating the effects of anharmonicity are not required.

When evaluating the numerator in the RRKM rate equation, $N^{\ddagger}(E - E_0)$, the vibrational modes which are conserved when progressing from the reactants to the products, need to be separated from those disappearing vibrational modes of the reactants which are converted into rotational and translation modes in the products. This can be visualized with the following example; a reactant ion with five atoms may have nine vibrational modes. If one bond breaks then one mode is lost, as it becomes the reaction coordinate. This leaves the transition state with eight vibrational modes, and the product four-atom daughter ion may have only six vibrational modes. These six vibrational modes are conserved, of the others, one is the reaction coordinate and the remaining two modes are the disappearing transitional modes. The transitional modes are easily identifiable in the transition state

1.3 The Study of Ionic Dissociations

as those which are altered considerably from the vibrational modes of the reactant molecule, and typically have low frequencies which can be tens or hundreds times smaller than the conserved frequencies. How these transitional modes are accounted for when calculating $N^{\ddagger}(E - E_0)$ gives rise to the different variations of the original RRKM rate theory.

In rigid-activated-complex (RAC- RRKM), the transitional modes are treated as harmonic oscillators and the transition state is fixed. In other words, the transition state is well-defined, possessing all the vibrational modes of the reactant molecule *minus* the vibrational mode which becomes the reaction coordinate taking the reaction through to the products [52]. This method adequately describes the rate curves for reactions with reverse barrier where the transition state structure is located upon the barrier, especially when a rearrangement is involved (see Chap. 5). As a consequence of this early transition barrier where the ion is still whole, the transition state is tighter and so the vibrational frequencies tend to be higher than for the equilibrium ion geometry. This produces a rate curve which is less steep than with other methods, as the energy dependence of the number of states is reduced [58].

However, this method is less suited to ionic dissociative photoionization reactions without a well-defined transition state and hence no reverse barrier, especially if the kinetic shift is large [58, 62]. For larger ions with many low energy modes, the minimum rate constants are typically well below 10^2 s^{-1} and the rate constants need to be accurate over several orders of magnitude to provide a reliable extrapolation down to E_0 [58]. This can lead to an underestimation of the onset at low ion internal energies while simultaneously failing to fit the rates at higher energies [63].

Phase space theory (PST) is at the opposite end of the scale to RAC-RRKM theory. In PST, energy and angular momentum are conserved. PST seeks to address the shortcomings of making the assumption that the translational of products results from the kinetic energy along the reaction coordinate, by considering the centrifugal barriers. The transitional modes are treated as free rotations and the product vibrational frequencies *and* moments of inertia are used in the calculation of the number of states. In the absence of a reverse barrier, the transition state is positioned at an infinite distance along the reaction coordinate. The assumption is made that *all* transitional modes are converted into barrierless rotational and translational modes of the products. This is visualized as an anharmonic potential energy curve where dissociation occurs at the asymptotic limit, and the transition state is located anywhere along increasing R distance at the asymptotic limit. The rate is therefore determined by the phase space available to those products [62, 64–66]. PST can be considered as equivalent to RAC-RRKM when the transition state structure is sufficiently loose [65]. However, at higher energies this method tends to overestimate the rates because the anisotropy of the system is not considered. Consequently this method is best reserved for low energies close to the onset, as the reduced number of vibrational modes (i.e. none) means the energy dependence upon the number of states is significant.

In the simplified statistical adiabatic channel model (SSACM) the number of states of the transitional modes are calculated as in phase space theory (PST) and then scaled with an energy-dependant 'rigidity' factor (tightness of the transition state) which prevents the rate constant from rising too rapidly at higher ion internal energies [47, 58, 62]. Unlike the previous methods, anisotropy of the potential energy surface is accounted for by the inclusion of the rigidity factors because the factors are specific to the different types of potential energy surfaces which characterize the dissociation process [67]. The disadvantage of this method is that while found to be valid at low energies, it may give unphysical behaviour at higher ion internal energies giving a rate curve with a negative slope [68].

The final theory to be mentioned is variational transition state theory (VTST). This theory is centred on finding the entropic minimum. In the absence of a reverse barrier, this minimum is the effective transition state where the sum of states is at a minimum, and it can be found by locating the global minimum in the sum of states, $N^{\ddagger}(E - V(R))$, as the molecule travels across the potential energy surface $V(R)$, along the reaction coordinate, R. Essentially, at each ion internal energy along the reaction coordinate, the minimum of the number of states function is determined, and the corresponding number of states is used to calculate the dissociation rate [69, 70]. Two minima can be located corresponding to the tight transition state minima at smaller values of R and an orbiting state at larger values of R. In the presence of a reverse barrier, the tight transition state is located at the top of that barrier. The orbiting transition state (at large R) coincides with the centrifugal barrier near the products. Both minima move to shorter bond distances as the energy increases, and at a particular energy, a change from the orbital transition state being the global minimum to the tight transition state may occur. It is no longer required to supply the exact transition state structure because by minimizing the sum of states, the entropy bottleneck is found [70]. These transition states may vary with the energy of the system and angular momentum. It remains unclear however, if for an ionic dissociation whose potential energy surface has only one well, the two transition state entropic minima produced with this method, are physically meaningful [71].

1.3.4 Kinetic and Competitive Shifts

Kinetic shifts are observed when an ion dissociates slowly, on a timescale comparable to or longer than that of the experiment [72]. In other words, if the ion falls apart too slowly then even if the ion has enough energy to dissociate, there is a possibility that the ion will have reached the detector before dissociation occurs.

This shifts the onset to higher energies (Fig. 1.6), leading to a disappearance energy of the parent ion several hundred meV or even eV higher than the actual onset. The difference between the onset as perceived on the breakdown diagram where the parent ion signal reaches zero, and the actual onset, is termed the kinetic shift, and is purely a consequence of the experiment dimensions. The shift arises because reactions with a reverse barrier tend to have tight transition states, producing the slow rates. For larger molecules, other pathways such as infrared emission from vibrationally excited ion states can compete with dissociation at low energies,

1.3 The Study of Ionic Dissociations

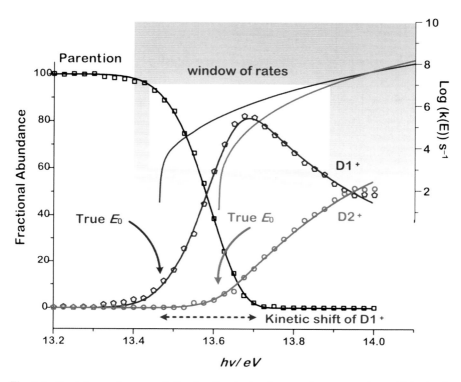

Fig. 1.6 Breakdown diagram of the fractional abundances of a parent ion which slowly dissociates into daughter ion 1 (D1$^+$), and the parallel dissociation forming daughter ion 2 (D2$^+$), as a function of photon energy, $h\nu$. The E_0 of for fast dissociations into D1$^+$ is given usually where the parent ion signal disappears, but in this instance, extrapolation gives the true E_0 at a lower energy. The time range of which rates can be experimentally measured is the 'window of rates', slow rates are below $k(E)$ 10^3 s^{-1}, fast rates are above $k(E)$ 10^7 s^{-1}

further obscuring the onset [63]. The true E_0 is then obtained by modelling the rates using the unimolecular rate theories described in the preceding section [73].

A *competitive shift* arises when another fragmentation channel, with comparable rates to the existing reactions, also becomes accessible at similar energies. The onset of this new channel is not observed at the dissociation threshold, but at a higher energy, only when it is of a similar rate with the fastest reaction. How quickly this competing channel catches up with the other fast reactions depends on its transition state frequencies. This is not the same as saying that the *rate* of one reaction is affected by the presence of another reaction channel [37].

1.4 Thermochemistry

1.4.1 Thermochemical Values Derived from iPEPICO

In the absence of an overall reverse barrier, the onset energy, E_0, is equivalent to the enthalpy of reaction at 0 K, $\delta_r H^\ominus_{0K}$. This is related to the enthalpies of formation by the following reaction:-

$$E_0 = \Delta_r H^\theta_{0\,textrmK} = \Delta_f H^\theta_{0K}[ion] + \Delta_f H^\theta_{0K}[neutral fragment] \\ - \Delta_f H^\theta_{0K}[neutral\ parent\ molecule] \quad (1.12)$$

Hence, if two out of the three energies are well established, then the third, less well determined value may be found. This simple relationship is used to great effect to determine unknown and difficult to establish (using purely ab initio methods) enthalpies of formation, in Chaps. 4 and 5. It is important to note, that thermochemical values may only be extracted from E_0 values derived from reactions where there is no reverse barrier to the products. Extrapolation to E_0 from slow dissociations yields a greater uncertainty than from fast dissociations, and thermochemical information derived from slow reactions is also less accurate.

To convert the 0 K enthalpy of formation of a molecular species to that at 298 K the following relationship is used,

$$\Delta_f H^\theta_{0K} + \sum \left(H^\circ_{298\,K} - H^\circ_{0K}\right)_{molecule} - \sum \left(H^\circ_{298\,K} - H^\circ_{0K}\right)_{constituent\ elements} \\ = \Delta_f H^\theta_{298\,K} \quad (1.13)$$

where the thermal correction term for a non-linear molecular species is defined as,

$$\left(H^\circ_{298\,K} - H^\circ_{0\,K}\right) \simeq \frac{5}{2} k_B T + \frac{3}{2} k_B T + \sum_{vib} \frac{h\nu}{\exp\left(\frac{h\nu}{k_B T}\right) - 1} \\ = 4 k_B T + \sum_{vib} \frac{h\nu}{\exp\left(\frac{h\nu}{k_B T}\right) - 1} \quad (1.14)$$

where $k_B T$ is the Boltzmann constant. The enthalpy of reaction at 298 K is then given by the difference between the sum of the enthalpies of formation of the products, and that of the reactants at 298 K.

For ionization reactions producing cations or anions, there are two different conventions as to how the electron is treated. In the Lias compendium [74], the *stationary* electron (or ion) convention is used where the electron is treated as a sub-atomic particle. Other values for charged species, such as those found in the JANAF (Joint Army Navy Airforce) compilation [75], the Burcat [76] and the Active Thermochemical Tables [77], use the *thermal* electron convention where the electron is defined as a standard chemical element. As such, there is a discrepancy in the enthalpy of formation of a cation or anion between these two conventions of 2.5 RT, or 6.20 kJ mol^{-1} at 298 K. For cations, the thermal convention values are the more positive. Equation 1.15 shows this contribution, where an electron is treated as an ideal gas following Boltzmann statistics, given the heat capacity of the electron, C_p (electron). The inclusion of +6.20 kJ mol^{-1} is only required for positive ionic species that are not at 0 K.

$$\int_{0\,K}^{298\,K} C_p(electron)dT = \left(H°_{298\,K} + H°_{0\,K}\right)_{electron} = \frac{5}{2}k_B T \text{ per electron}$$
$$= 6.20 \text{ kJmol}^{-1} \text{ at 298 K} \quad (1.15)$$

This point is important, as in some instances comparisons are made between values determined in this work and those from the different compilations. Throughout this work, e.g. in Chaps 4 and 5, the values for $\Delta_f H^\ominus$ of cations at temperatures other than at 0 K use the stationary (ion) convention, in line with the Lias tables. Enthalpies of reaction can be used to identify whether a reaction is energetically feasible. In Chap. 5, multiple photodissociation reactions produce ion fragments of the same mass, and these thermochemical onsets are used successfully to deduce which reaction gives rise to the observed signal. Determining enthalpies of formation also reveals which dissociation reactions are energetically allowed but not observed and are therefore blocked on the potential energy surface. The difference between the calculated $\Delta_r H^\ominus_{0K}$ values from the experimental onset can be used to gauge the magnitude of any reverse barriers, the *competitive shift* and identify if tunnelling through the potential energy barrier occurs.

1.4.2 Isodesmic Reactions

Isodesmic reaction energy calculations feature among the ab initio and experimental tools to determine thermochemical properties. What is an isodesmic reaction, and how do they fit in with determining enthalpies of formation? Isodesmic, derived from the Greek *isos* for equal and *desmos* for bond [78] means a reaction where, there are the same numbers of the same type of bonds on both the reactants *and* products side of the equation. An example using fluorinated ethenes is given below,

$$F_2C = CF_2 + H_2C = CF_2 \rightarrow 2\text{ HFC} = CF_2 \quad (1e)$$

Isodesmic reactions were initially developed in order to combat the neglect of electron correlation effects inherent in computationally inexpensive ab intio methods, giving rise to underestimated dissociation energies. They have been used successfully to predict the thermochemistry of a range of systems [78, 79]. Typically, ab initio enthalpies of reaction are used together with experimentally determined enthalpies of formation to determine an unknown enthalpy of formation. Ab initio techniques have often been found to be robust and proficient methods of calculating enthalpies of formation; however they do so at considerable computational cost. Therefore, this particular route may not always be ideal as the size of the examined system increases; instead a much quicker and simpler method is required. The main incentive of using isodesmic reactions is their reduced computational cost and the balancing of systematic errors inherent in ab initio calculations, effectively cancelling their effects [76].

Within the umbrella of isodesmic reactions, there is a hierarchy of related reactions. The simplest of these is the *isogyric* reaction which only the number of electron pairs are conserved and the products are methane molecules, with molecular hydrogen used to balance the reactant side of the equation. *Isodesmic* reactions preserve the quantity of identical bonds between reactants and products, and the reaction is balanced with the addition of the appropriate number of parent molecules. *(Hypo) homodesmotic* reactions are defined as those having equal numbers of carbon atoms in their specific modes of hybridization in both reactants and products, and equal numbers of the same types of carbon–hydrogen bonds [80]. *(Hyper) homodesmotic* reactions are a subset of (Hypo) homodesmotic reactions, and are equations in which there are *equal numbers of carbon–carbon bond* types inclusive of carbon hybridization and number of hydrogens attached [81, 82]. The increasing detail accommodated by the range of reactions makes their use a suitable approach to determine thermochemical properties for a comprehensive collection of molecules. They can be used for systems containing single bonds, to conjugated systems with greater numbers of multiple bonds [82, 83]. Isodesmic reactions have proved to be a useful tool in studying closed shell systems [62, 84, 85]. In Chap. 4, isoelectronic reactions, a variation of isodesmic reactions, using enthalpies of formation of small ions are used. Whilst the differing systematic errors inherent in the primary ab initio calculations are not so effectively cancelled as is the case for closed shell systems, they do not seem to affect the overall result greatly for small systems [82]. Isodesmic reactions provide a compromise between purely calculated values and those experimentally derived [85]. They are used extensively throughout Chap. 4 to determine the hitherto ill-defined enthalpies of formation of bromine containing molecules and ions.

References

1. Simpson, M. J. (2012). *Two studies in gas-phase ion spectroscopy, vacuum-ultraviolet negative photoion spectroscopy and ion-molecule reaction kinetics*. Berlin: Springer.
2. Eland, J. H. D. (1984). *Photoelectron spectroscopy*. London: Butterworths.
3. Hollas, J. M. (1998). *High resolution photoelectron spectroscopy* (2nd ed.). Chichester: Wiley & Sons Ltd.
4. Softly, T. P. (2004). *International Reviews in Physical Chemistry, 23*, 1–78.
5. Chupka, W. A., Miller, P. J., & Eyler, E. E. J. (1988). *Chemical Physics, 88*, 3032–3036.
6. Guyon, P. M., Baer, T., & Nenner, I. J. (1983). *Chemical Physics, 78*, 3665–3672.
7. Jensen, F. (2007). *Introduction to Computational Chemistry* (2nd Edition). John Wiley & Sons: Chichester.

8. Atkins, P., & Friedman, R. (2005). *Molecular quantum mechanics*. Oxford: Oxford Univeristy Press.
9. Koopmans, T. (1934). *Physica, 1*, 104–113.
10. Roothan, C. C. J. (1951). *Reviews of Modern Physics, 23*, 69–89.
11. Landau, A., Khistyaev, K., Dolgikh, S., & Krylov, A. I. J. (2010). *Chemical Physics, 132*, 014109.
12. Stowasser, R., & Hoffmann, R. J. (1999). *American Chemical Society, 121*, 3414–3420.
13. Baerends, E. J., & Gritsenko, O. V. J. (1997). *Physical Chemistry A, 101*, 5383–5403.
14. Condon, E. (1926). *Physical Review, 28*, 1182–1201.
15. Ellis, A., Feher, M., & Wright, T. (2005). *Electronic and photoelectron spectroscopy*. Cambridge: Cambridge Univeristy Press.
16. Born, M., & Oppenheimer, J. R. (1927). *Annalen der Physik (Leipzig), 84*, 457–484.
17. Levine, R. D. (2005). *Molecular reaction dynamics*. Cambridge: Cambridge Univeristy Press.
18. Worth, G. A., & Cederbaum, L. S. (2004). *Annual Review of Physical Chemistry, 55*, 127–158.
19. Köppel, H., Cederbaurn, L. S., Domcke, W., & Shaik, S. S. (1983). *Angewandte Chemie International Edition in English, 22*, 210–224.
20. Köppel, H., Cederbaum, L. S., & Domcke, W. J. (1982). *Chemical Physics, 77*, 2014–2022.
21. Sannen, C., Raşeev, G., Galloy, C., Fauville, G., & Lorquet, J. C. J. (1981). *Chemical Physics, 74*, 2402–2412.
22. Willitsch, S., Hollenstein, U., & Merkt, F. J. (2004). *Chemical Physics, 120*, 1761–1774.
23. Bernardi, F., Olivucci, M., & Robb, M. A. (1996). *Chemical Society Reviews, 25*, 321–328.
24. Illenberger, E., & Momigny, J. (1992). *Gaseous molecular ions, an introduction to elementary processes induced by ionization*. New York: Springer.
25. Werner, A. S., Tsai, B. P., & Baer, T. (1974). *Journal of Chemical Physics, 60*, 3650–3657.
26. Lifshitz, C., & Long, F. A. J. (1963). *Physical Chemistry, 67*, 2463–2468.
27. Lifshitz, C., & Gleitman, Y. (1981). *International Journal of Mass Spectrometry and Ion Physics, 40*, 17–29.
28. Puttkammer, E. v. Z. *Naturforsch. A*. (1970). *25*, 1062–1071.
29. Baer, T. (2000). *International Journal of Mass Spectrometry, 200*, 443–457.
30. Powis, I. (1983). *Chemical Physics, 74*, 421–432.
31. Baer, T., Booze, J. A., & Weitzel, K. M. (1991). *Vacuum ultraviolet photoionization and photodissociation of molecules and clusters*. Singapore: World Scientific.
32. Eland, J. H. D. (1972). *International Journal of Mass Spectrometry and Ion Processes, 9*, 397.
33. Merkt, F. (1997). *Annual Review of Physical Chemistry, 48*, 675–709.
34. Dressler, R. A., Chiu, Y., Levandier, D. J., Tang, X. N., Hou, Y., Chang, C., et al. (2006). *Chemical Physics, 125*, 132306–132451.
35. Ng, C. J. (2000). *Electron Spectroscopy and Related Phenomena, 112*, 31–46.
36. Jarvis, G. K., Weitzel, K.-M., Malow, M., Baer, T., Song, Y., & Ng, C. (1999). *Review of Scientific Instruments, 70*, 3892–3906.
37. Baer, T., Sztáray, B., Kercher, J. P., Lago, A. F., Bodi, A., Skull, C., et al. (2005). *Physical Chemistry Chemical Physics (PCCP), 7*, 1507–1513.
38. Qian, X.-M., Lau, K.-C., He, G. Z., Ng, C. Y., & Hochlaf, M. J. (2004). *Chemical Physics, 120*, 8476–8485.
39. Weitzel, K.-M., Malow, M., Jarvis, G. K., Baer, T., Song, Y., & Ng, C. Y. J. (1999). *Chemical Physics, 111*, 8267–8270.
40. Chupka, W. A. J. (1993). *Chemical Physics, 98*, 4520–4531.
41. Leach, S., Devoret, M., & Eland, J. H. D. (1978). *Chemical Physics, 33*, 113–121.
42. Klosterjensen, E., Maier, J. P., Marthaler, O., & Mohraz, M. J. (1979). *Chemical Physics, 71*, 3124–3128.
43. Ibukia, T., Shimadaa, Y., Hashimotoa, R., Nagaokab, S., Hinob, M., Okadac, K., et al. (2005). *Chemical Physics, 314*, 119–126.
44. Molloy, R. D., & Eland, J. H. D. (2006). *Chemical Physics Letters, 421*, 31–35.

45. Eland, J. H. D. J. (2000). *Electron Spectroscopy and Related Phenomena, 112,* 1–8.
46. Eland, J. H. D., Fink, R. F., Linusson, P., Hedin, L., Plogmakerd, S., & Feifeld, R. (2011). *Physical Chemistry Chemical Physics (PCCP), 13,* 18428–18435.
47. Sztáray, B., Bodi, A., & Baer, T. J. (2010). *Mass Spectrometry, 45,* 1233–1245.
48. Borkar, S., & Sztáray, B. J. (2010). *Journal of Physical Chemistry A, 114,* 6117–6123.
49. Kercher, J. P., Stevens, W., Gengeliczki, Z., & Baer, T. (2007). *International Journal of Mass Spectrometry, 267,* 159–166.
50. Bodi, A., Shuman, N. S., & Baer, T. (2009). *Physical Chemistry Chemical Physics (PCCP), 11,* 11013–11021.
51. Lifshitz, C. J. (1983). *Physical Chemistry, 87,* 2304–2313.
52. Beynon, J. H., & Gilbert, J. R. (1984). *Application of Transition State Theory to Unimolecular Reactions.* John Wiley & Sons Ltd: Chichester.
53. Simm, I. G., Danby, C. J., Eland, J. H. D., & Mansell, P. I. (1976). *Journal of the Chemical Society, Faraday Transactions, 2*(72), 426–434.
54. Smith, D. M., Tuckett, R. P., Yoxall, K. R., Codling, K., & Hatherly, P. A. (1993). *Chemical Physics Letters, 216,* 493–502.
55. Marcus, R. A., & Rice, O. K. J. (1951). *Phys. Coll. Chem., 55,* 894–908.
56. Baer, T., & Hase, W. L. (1996). *Unimolecular reaction dynamics: theory and experiments.* New York: Oxford University Press Inc.
57. Baer, T., & Mayer, P. M. J. (1997). *The Journal of the American Society for Mass Spectrometry, 8,* 103–115.
58. Stevens, W., Sztáray, B., Shuman, N. S., Baer, T., & Troe, J. J. (2009). *Journal of Physical Chemistry A, 113,* 573–582.
59. Beyer, T., & Swinehart, D. R. (1973). *ACM Communications, 16,* 379.
60. Whitten, G. Z., & Rabinovitch, B. S. J. (1963). *Chemical Physics, 38,* 2466–2474.
61. Forst, W. (1973). *Theory of unimolecular reactions.* New York: Academic.
62. Troe, J., Ushakov, V. G., & Viggiano, A. A. J. (2006). *The Journal of Physical Chemistry A, 110,* 1491–1499.
63. Shuman, N. S., Bodi, A., & Baer, T. J. (2010). *The Journal of Physical Chemistry A, 114,* 232–240.
64. Light, J. C. J. (1964). *Chemical Physics, 40,* 3221–3229.
65. Klots, C. E. J. (1971). *Physical Chemistry, 75,* 1526–1532.
66. Chesnavich, W. J., & Bowers, M. T. J. (1977). *Chemical Physics, 66,* 2306–2315.
67. Troe, J. J. (1997). *Journal of the Chemical Society, Faraday Transactions, 93,* 885–891.
68. Shuman, N. S., Miller, T. M., Viggiano, A. A., & Troe, J. J. (2011). *Chemical Physics, 134,* 094310–094320.
69. Chesnavich, W. J. J. (1986). *Chemical Physics, 84,* 2615–2619.
70. Chesnavich, W. J., Bass, L., Su, T., & Bowers, M. T. J. (1981). *Chemical Physics, 74,* 2228–2246.
71. Hu, X., & Hase, W. L. J. (1989). *Phys. Chem., 93,* 6029–6038.
72. Lifshitz, C. (1982). *Mass Spectrometry Reviews, 1,* 309–348.
73. Chupka, W. A. J. (1959). *Chemical Physics, 30,* 191–211.
74. Lias, S. G., Bartmess, J. E., Liebman, J. F., Holmes, J. L., Levin, R. D., & Mallard, W. G. (1988). *Journal of Physical and Chemical Reference Data, 17,* 872.
75. Lias, S. G. (2011). "Ionization Energy Evaluation". ; In P. J. Linstrom & W. G. Mallard (Eds.), In *NIST Chemistry WebBook, NIST Standard Reference Database Number 69* (p. 20899). Gaithersburg: National Institute of Standards and Technology.
76. Burcat, A., & Ruscic, B.Sesptember 2005 third millennium ideal gas and condensed phase thermochemical database for combustion with updates from active thermochemical Tables, ANL-05/20 and TAE 960, Technion-IIT; Aerospace Engineering, and Argonne National Laboratory, Chemistry Division September 2011. ftp://ftp.technion.ac.il/pub/supported/aetdd/thermodynamics mirrored at http://garfield.chem.elte.hu/Burcat/burcat.html.
77. Ruscic, B. Active Thermochemical Tables, early beta 1.110, http://atct.anl.gov/index.html, May 09 2012.

78. Hehre, W. J., Ditchfield, R., Radom, L., & Pople, J. A. J. (1970). *Journal of the American Chemical Society, 92,* 4796–4801.
79. Radom, L., Hehre, W. J., & Pople, J. A. J. (1971). *Journal of the American Chemical Society, 93,* 289–300.
80. George, P., Trachtman, M., Bock, C. W., & Brett, A. M. (1975). *Theoretica Chimica Acta, 38,* 121–129.
81. George, P., Trachtman, M., Bock, C. W., & Brett, A. M. (1976). *Journal of the Chemical Society, Perkin Transactions, 2,* 1222–1227.
82. Wodrich, M. D., Corminboeuf, C., & Wheeler, S. E. J. (2012). *Journal of Physical Chemistry A, 116,* 3436–3447.
83. Wheeler, S. E., Houk, K. N., Schleyer, P. V. R., & Allen, W. D. (2009). *Journal of the American Chemical Society , 131,* 2547–2560.
84. Bodi, A., Kvaran, Á., & Sztáray, B. J. (2011). *Journal of Physical Chemistry A, 115,* 13443–13451.
85. Bodi, A., Hemberger, P., & Gerber, T. J. (2013). *Chemical Thermodynamics, 58,* 292–299.

Chapter 2
Experimental

2.1 Preamble

This chapter concerns the experimental facet of the work undertaken during this PhD. It begins by describing the light source used to ionize the molecules, followed by how that light is manipulated and a description of the *endstation* with which the experiments are performed. It concludes by detailing how the electron and ion signals are detected, and finally how the data is prepared for analysis.

The imaging photoelectron photoion coincidence (iPEPICO) apparatus, which was used throughout this work to record both the coincidence results and the threshold photoelectron spectra (TPES), is located at the vacuum ultraviolet beamline of the Swiss Light Source (SLS), a third generation synchrotron light source. It has been in operation since 2008 to investigate a range of systems from small gas phase molecules [1–3], to radicals produced by photolysis [4], and larger systems such as paracyclophanes [5] and organometallic compounds [6, 7]. The beamline and endstation have been described in detail in several publications [8–11].

2.2 The Synchrotron Radiation Source

A radiation source is required to excite electrons in the valence shell and ionize the sample creating ionic and neutral products, the former of which are measured in the iPEPICO experiment. One such radiation source is synchrotron light, which is produced in particle accelerators by electrons moving at relativistic speeds through magnetic fields. The accelerators are generally circular and are comprised of an electron source (electron gun), a linear accelerator (Linac), a booster ring that brings the energy of the electron bunches up to their final energy, at the SLS this is 2.4 GeV, and a storage ring, which in the case of the SLS has a diameter of 288 m. The SLS was the first synchrotron to operate in a top-up injection mode; meaning that the current of electron bunches travelling around the ring is kept almost constant at 400 mA [12]. This is in contrast to many other synchrotrons where the electron current decays with time. When the stream of electron bunches is deflected round the corners of the storage ring (facilitated by bending magnets), they emit light over a range of wavelengths. Due to relativistic effects this high intensity electromagnetic radiation is emitted in a highly collimated beam. The high flux radiation therefore possesses a small divergence (high brilliance), is polarized and emitted across the entire spectrum. Insertion devices integrated in the storage ring, such as undulators and wigglers, alter the path of the electrons to generate even more tuneable and intense radiation in the straight sections of the storage ring [13]. The radiation leaves the ring tangentially via *beamlines*. The light delivered into the vacuum ultra violet (VUV) beamline is provided by a bending magnet, and is linearly polarized in the plane of the storage ring but elliptically polarized above and below the plane [9].

There are other means of supplying the required ultraviolet and vacuum ultraviolet wavelengths that include laboratory based discharge lamps or laser sources. However, for performing TPEPICO and TPES experiments, they all have weaknesses compared to synchrotron radiation. Easily constructed discharge lamps, e.g. the non-tuneable Helium I discharge lamps which emit the He Iα resonance line at 21.22 eV, corresponding to the He* $^1P(1s^1\ 2p^1)$ to He $^1S(1s^2)$ transition have been used to great effect in many works [14–19]. Another example is the Hydrogen many-line discharge lamp [20], which offers tuneability across the range from 8.75 to 13.8 eV. However, all discharge lamps tend to have low brilliance, meaning lower ionization rates compared with synchrotron light, making many experiments time consuming. Laser systems offer brilliant, coherent light. For example, the Nd:YAG (Nd^{3+} ions doped in a rod of yttrium aluminium garnet) provide high coherent photon flux e.g. 10^{17} photons per second and $E/\Delta E > 10^7$ together with low divergence, high monochromaticity and the opportunity to control the dynamics of a system [21, 22]. These lasers are often used to pump dye lasers which are typically used as the exciting photon source. Dye lasers offer tuneability into the near ultraviolet and vacuum ultraviolet regions which is achieved through frequency doubling of the initial wavelengths, but this can be a complex process. In contrast to these laboratory based light sources, light produced from a synchrotron source encompasses a large range of wavelengths (X-rays to infra-red, 10^{-10}–10^{-5} m), is plane-polarized, bright, pseudo-continuous-wave and easily tuneable using a diffraction grating. The intensity of light delivered through the beamline was measured to be approximately $10^{11}\ s^{-1}$. Therefore the light intensity or ionization rate is many times than at the Lyman-α line at 1215.67 Å or 10.198 eV, with a conventional tuneable hydrogen discharge lamp [8].

At the beginning of the VUV *beamline*, radiation delivered by the bending magnet across the vacuum ultraviolet wavelengths is selected with a monochromator. The *beamline* is kept under high vacuum (10^{-10}–10^{-9} mbar), which is essential to maintain the high vacuum in the storage ring and prevent carbon deposits on the optical elements. Furthermore as oxygen absorbs in the VUV region, from 185 nm to low wavelength a high vacuum is required to maintain transmittance of the VUV radiation through to the experiment [8]. Two laminar gratings ruled with 600 and 1200 lines/mm which can be interchanged under vacuum [23], are used to access the range between 5 and 30 eV [9]. Higher harmonics of radiation occur when radiation of double the energy, e.g. at 20 eV (second harmonic), is transmitted from the grating together with that of the first harmonic e.g. at 10 eV. Therefore a spectrum recorded between 10 and 15 eV will be contaminated by additional structure usually seen at 20–30 eV. Below 11 eV, higher harmonics of the radiation can be removed using a MgF_2 window which virtually absorbs all light above 11 eV. However no solid window suitably removes higher harmonics at higher energies. At the VUV beamline, this is done using a differentially pumped noble gas filter positioned in front of the iPEPICO endstation (Fig. 2.1.). Filtering is achieved because all photons with energy above the ionization limit of the gas are absorbed, while all those below are transmitted.

Different noble gases are used depending on the energy range being scanned, for example, 10 mbar of a Neon, Argon and Krypton mix can be used up to 14.0 eV and 8 mbar of Ne is used below 21.5 eV [8, 24]. No gas filter is used at higher photon energies. However the gases may absorb specific wavelengths of light corresponding to their electronic absorptions. These absorptions can contaminate the experimental spectra, are sharp and well defined, and in most cases they can simply be removed from the spectrum without issue. However, special care is required when removing absorption lines which coincide with fine spectral features, such as vibrational progressions, so as not to lose spectral information.

Fig. 2.1 The compact differentially pumped noble gas filter, which removes higher harmonics of radiation

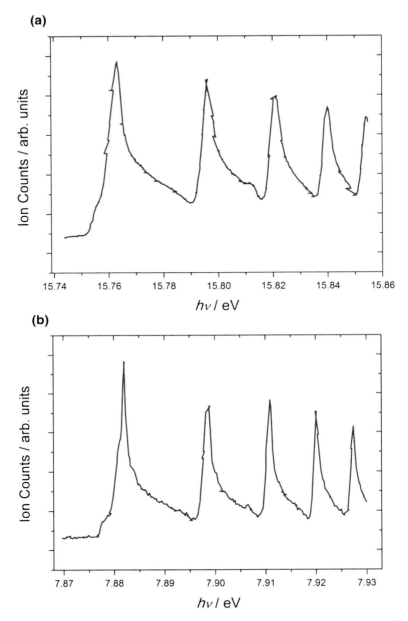

Fig. 2.2 **a** The ion yield of $^2P_{3/2}$ Ar threshold (first harmonic). **b** The ion yield of $^2P_{3/2}$ Ar threshold (second harmonic)

2.2 The Synchrotron Radiation Source

The photon energy is calibrated against Argon 11s′, 15.7639 eV, 12s′, 15.7973 eV, 11d′, 15.8210 and 14s′, 15.84047 eV autoionization peaks, and with those given by the second harmonic at half the energy (see Fig. 2.2) [25]. The photon energy resolution is 3 meV at 10 eV.

2.3 The Endstation

The endstation, pictured in Fig. 2.3, was constructed by Dr. Andras Bodi et al. [9], in situ at the SLS. The pure sample is introduced into the experimental chamber through an effusive source at room temperature, with typical pressures in the chamber of $2-4 \cdot 10^{-6}$ mbar during measurement. The background pressure is in the order of 10^{-7} mbar. Low operating pressures are necessary because the ions and ejected electrons must be allowed to travel to their respective detectors unimpeded by background sample gas. The sample is ionized by the incident monochromatic VUV synchrotron radiation dispersed by a grazing incidence monochromator [8].

Following photoionization, the photoelectrons and photoions are accelerated in opposite directions by a constant extraction field of 20–120 V cm^{-1}. The ejected electrons are velocity map imaged (VMI) onto the 40 × 40 mm DLD40 Roentdek position sensitive delay-line detector, whereby threshold electrons are focused using electrostatic lenses onto a smaller than 1 mm spot on the centre of with a kinetic energy resolution of 1 meV at threshold. The mainstay of VMI is that electrons with the *same* initial velocity (and by consequence, the same kinetic energy) are focussed onto the same area on the detector regardless of where in the ionization region they were formed [26]. Energetic electrons are detected as rings

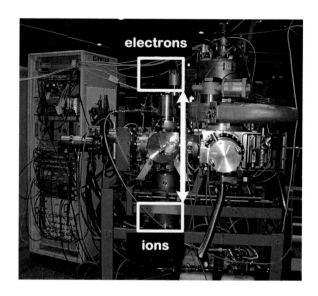

Fig. 2.3 The iPEPICO experiment. VUV light enters from behind, perpendicular to the electorn/ion flight axis (*double headed arrow*). The electron signal is recorded with a Roendek DLD40 delay-line-detector, and the ions are recorded using a Jordan C-726 detector

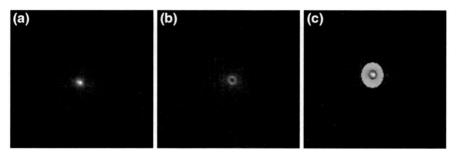

Fig. 2.4 The electron signal image of 1,1-$C_2H_2F_2$ at 15.325 eV, as captured by the DLD40 position sensitive detector. **a** The image, **b** the *circle* area where threshold electron signal is located, highlighted *blue* and **c** the *ring* area highlighted *yellow* where the background (*hot*) electrons are

around the central threshold electron circle (see Fig. 2.4.), where the radii is proportional to their initial velocity perpendicular to the extraction axis. Using electrostatic plates to focus the electrons prevents distortion of the image, such as blurring that occurs when imaging using gridded extraction plates. With grids, image quality is restricted by the size of the ionization volume [27]. With VMI, threshold signal can be obtained from a larger ionization area, without sacrificing signal quality [26].

After acceleration in the 5 cm long 120 V cm^{-1} primary acceleration region, the ions undergo a further acceleration to −1800 V, which provides the necessary space focusing conditions [8]. This is achieved using the standard Wiley–McLaren conditions [28], where the arrival time of the ion is independent of where it was formed in the ionization region. This means that those ions formed at the top of the ionization region (towards the electron detector) are born at a more positive potential, ions formed further away, at the bottom of the ionization region towards the ion detector, at a more negative potential and reach the detector virtually simultaneously to give the lower and upper TOF distribution limit [15, 28]. A compromise is reached between electron energy resolution which decreases with increasing electric field, and ion mass resolution, which increases with increasing electric field. This is important because lower electric fields prolong the residence time of the dissociating ion in the acceleration region, thereby enabling the rate constants to be measured and kinetic shifts to be observed in instances of slowly (metastable) dissociating ions. This means that where necessary, as in the in the case of ions formed with large kinetic energy release giving rise to substantially broadened ion time-of-flight peaks, larger fields can be used to draw out the ions without sacrificing electron energy resolution. Ions then enter the 55 cm field free drift region and are finally detected by a Jordan TOF C-726 microchannel plate assembly [8].

2.4 Capturing the Electron and Ion Signals

All of the emitted electrons are focussed onto the imaging detector and as previously described; electrons with low kinetic energy are focussed onto the centre of the detector and electrons with electronic with significant kinetic energy (hot electrons) are focussed around this central spot. However some of the hot electrons have a velocity vector that is already oriented along the flight tube axis and also arrive at the centre of the detector, thereby contaminating the true threshold signal. As such, there needs to be some discrimination against hot electrons to yield the true threshold electron signal. One approach is to process the total signal to yield the threshold signal. Typically, the 3 dimensional expanding Newton sphere of charged particles is extracted from the 2 dimensional projection (image) as captured by VMI with techniques such as the onion peeling algorithm [29], inverse Abel transformations [8] or pBasex [30]. However noise can be generated in the reconstructed image for example, by over subtracting contributions made by the faster electrons.

In this instance, the hot electron contamination of the threshold signal is accounted for by a simple subtraction process, as introduced by Sztáray and Baer [31]. This method is preferred over the other techniques because it enables the use of high extraction fields to maximise ion TOF resolution without sacrificing the quality of the true threshold electron signal [31, 32]. The signal, as captured by the delay-line detector, from a small ring around the central spot, is subtracted from the central threshold signal (see Fig. 2.4). A greater signal to noise ratio, especially at higher energies where most electrons are very energetic providing an almost constant background, is produced when the threshold electrons are focused onto the small circle area, which is typically tenths of a mm in diameter (see Fig. 2.5). The true threshold signal is obtained by subtracting the background hot electron signal (multiplied by a factor which is the ratio of the two areas) from the threshold central spot signal. In the absence of a significant rotational envelope in the TPES, this subtraction process yields a true zero energy electron signal.

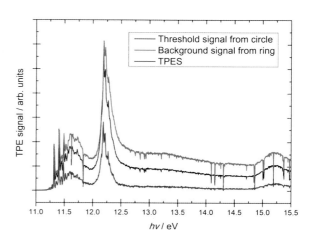

Fig. 2.5 The threshold electron signal of $CH_2Cl_2^+$ as captured by the *circle* area on the detector plotted with the background (*hot*) electron contamination as detected by the *ring* area and the final TPES. The *sharp lines* at 11.56, 11.76 eV and between 12.5 and 15.5 eV are absorption lines from the gas filter

Electron hit positions and times, and ion hits, are recorded using a time-to-digital converter card (HPTDC) operating in a special triggerless mode designed for the experiment [8]. This triggerless mode is designed to reduce the number of false coincidences recorded in the data and to improve the repetition of the experiment, to better correlate with the quasi-continuous nature of the synchrotron light. In any coincidence experiment, false coincidences will occur which are caused by the detection of an ion and an electron which were not formed in the same ionization event. These false coincidences can add considerable noise to the TOF spectrum reducing its quality.

Count rates supplied by data acquisition methods such as continuous single-start/single-stop (SS) and single-start/multiple-stop (SM) modes are not suitable for the high ionization rates produced with synchrotron based experiments. In (SS) data acquisition the TOF counting stops when the first stop signal is received, making it unsuitable for high intensity experiments. Start signals are lost while the time-to-amplitude converter (TAC) used to detect ion hits waits for the stop signal. After this initial stop signal, any additional stops are lost, so no discrimination between true and false coincidences is made [33] and the background signal drops exponentially from zero time [11, 34]. In single-start/multiple-stop (SM) [35, 36] counting starts after receiving the first start signal and all stops signals are recorded. Start signals are still lost while the acquisition cycle (using a time-to-digital converter) completes, and the false coincidence background is no longer constant for high event frequencies and electron collection efficiencies, and is therefore a limiting factor.

Instead, an alternative scheme is implemented in the iPEPICO apparatus. Time-of-flight ion distributions are obtained by correlating electrons and ions 'on the fly' with a multistart-multistop (MM) mode of data acquisition, reducing waiting (deadtime). All start signals are correlated with all stop signals within the relevant time signal, preventing paralysis of the ion or electron signal [15], to produce a constant background along the TOF spectrum [11]. (MM) mode of data acquisition is particularly suited to high intensity synchrotron work, as it enables data collection with the arbitrarily high ionization rates afforded by the intense synchrotron light. In pulsed experiments, ions accumulate in the ionization region until they are pulsed out to the detector which increases the number of false coincidences. Ion signals originating from false coincidences can be identified by their displacement from the ionization region and by their increased kinetic energy gained from the extraction pulse, thus appearing at an earlier TOF. Here, false parent ion coincidences can be distinguished, but it becomes less clear for false daughter ion signals which are produced with a distribution of translational energies [37]. The pulsed-extraction experimental setup is constrained by the need to minimize false coincidences whilst maximizing true coincidences, as the expected number of ionization events needs to be *circa* one per pulse, which considerably lengthens the data acquisition time. However, this does not apply with the continuous extraction employed in the iPEPICO set-up and count rates are limited by the false coincidence background signal.

2.5 The Experimental Results

The final results derived from the primary experimental data (as given by threshold electron and ion signals outlined above) are; the threshold ion TOF distributions measured as a function of photon energy, containing both the fractional ion abundances as well as the rate information in the form of asymmetric daughter ion peak shapes, and the threshold electron signal as a function of photon energy. The former can be concisely plotted in the breakdown diagram, i.e. the fractional ion abundances as a function of the photon energy, which includes most of the experimental information for fast dissociations. As the relative ratios of the ion abundances are plotted in the breakdown diagram, the TPEPICO technique is unaffected by changes in sample pressure, photon flux and varying Franck–Condon factors across photon energy. Plotting the threshold electron signal yields the threshold photoelectron spectrum (TPES). All electron counts were normalized to sample pressure (recorded with each point) and photon flux. The flux delivered to the endstation via the high and low energy diffraction grating, through the various noble gas filters, was measured by recording the fluorescence originating from the synchrotron light striking a sodium salicylate coated Pyrex window with photomultiplier tube.

References

1. Bodi, A., Kvaran, Á., & Sztáray, B. (2011). *Journal of Physical Chemistry A, 115*, 13443–13451.
2. Borkar, S., Sztáray, B., & Bodi, A. (2011). *Physical Chemistry Chemical Physics: PCCP, 13*, 13009–13020.
3. Bodi, A., Shuman, N. S., & Baer, T. (2009). *Physical Chemistry Chemical Physics: PCCP, 11*, 11013.
4. Hemberger, P., Noller, B., Steinbauer, M., Fischer, I., Alcaraz, C., Cunha de Miranda, B. R. K., et al. (2010). *Journal of Physical Chemistry A, 114*, 11269–11276.
5. Hemberger, P., Bodi, A., Schon, C., Steinbauer, M., Fischer, K. H., Kaiserc, C., et al. (2012). *Physical Chemistry Chemical Physics: PCCP, 14*, 11920–11929.
6. Pongor, C. I., Szepes, L., Basi, R., Bodi, A., & Sztáray, B. (2012). L. *Organomet., 31*, 3620–3627.
7. Sztáray, B. L., & Baer, T. (2002). *Journal of Physical Chemistry A, 106*, 8046–8053.
8. Bodi, A., Johnson, M., Gerber, T., Gengeliczki, Z., Sztáray, B., & Baer, T. (2009). *Review of Scientific Instruments, 80*, 034101.
9. Johnson, M., Bodi, A., Schulz, L., & Gerber, T. (2009). *Nuclear Instruments and Methods in Physical Research Section A, 610*, 597–603.
10. Sztáray, B., Bodi, A., & Baer, T. (2010). *Journal of Mass Spectrometry, 45*, 1233–1245.
11. Bodi, A., Sztáray, B., Baer, T., Gerber, T., & Johnson, M. (2007). *Review of Scientific Instruments, 78*, 084102.
12. http://www.psi.ch/sls/about-sls. Accessed 10/03/2012.
13. Hulbert, S. L., & Williams, G. P. (2000). *Vacuum Ultraviolet Spectroscopy*. London: Academic Press.

14. Dannacher, J., Schmelzer, A., Stadelmann, J.-P., & Vogt, J. (1979). *International Journal of Mass Spectrometry and Ion Physics, 31*, 175–186.
15. Eland, J. H. D. (1972). *International Journal of Mass Spectrometry and Ion Processes, 9*, 151–397.
16. Güthe, F., Locht, R., Leyh, B., Baumgärtel, H., & Weitzel, K.-M. (1999). *Journal of Physical Chemistry A, 103*, 8404–8412.
17. Locht, R., Leyh, B., Dehareng, D., Hottmann, K., & Baumgärtel, H. (2010). *Journal of Physics B: Atomic, Molecular and Optical Physics, 43*, 015102–015117.
18. Powis, I. (1983). *Chemical Physics, 74*, 421–432.
19. Locht, R., Dehareng, D., & Leyh, B. (2012). *Journal of Physics B: Atomic, Molecular and Optical Physics, 45*, 115101–115118.
20. Werner, A. S., Tsai, B. P., & Baer, T. (1974). *J. Chem. Phys, 60*, 3650–3657.
21. Ashfold, M. N. R., Nahler, N. H., Orr-Ewing, A. J., Vieuxmaire, O., Toomes, P. J., Kitsopoulos, R. L., et al. (2005). *Physical Chemistry Chemical Physics: PCCP, 8*, 26–53.
22. Penfold, T. J., Spesyvtsev, R., Kirkby, O. M., Minns, R. S., Parker, D. S. N., Fielding, H. H., et al. (2012). *Journal of Chemical Physics, 137*, 204310–204322.
23. Bodi, A., Hemberger, P., Gerber, T., & Sztáray, B. (2012). *Review of Scientific Instruments, 83*, 083105–083113.
24. Minnhagen, L., Palmeri, P., & Biémont, E. (2012). NIST Atomic Spectra Database (ver. 5.0). In http://physics.nist.gov/cgi-bin/ASD/energy1.pl. National Institute of Standards and Technology.
25. Ralchenku, Y., Kramida, A. E., & Reader, J. (2008). NIST ASD Team, Atomic Spectra Database, NIST, National Institute of Standards and Technology Gaithersburg, MD http://physics.nist.gov/asd3.
26. Eppink, A. T. J. B., & Parker, D. H. (1997). *Review of Scientific Instruments, 68*, 3477–3485.
27. Chandler, D. W., & Houston, P. L. (1987). *Journal of Chemical Physics, 87*, 1445–1447.
28. Wiley, W. C., & McLaren, I. H. (1955). *Review of Scientific Instruments, 26*, 1150–1158.
29. Manzhos, S., & Loock, H.-P. (2003). *Computer Physics Communications., 154*, 76–87.
30. Garcia, G. A., Nahon, L., & Powis, I. (2004). *Review of Scientific Instruments, 75*, 4989–4996.
31. Sztáray, B., & Baer, T. (2003). *Review of Scientific Instruments, 74*, 3763–3768.
32. Li, Y., Sztáray, B. L., & Baer, T. (2001). *Journal of the American Chemical Society, 123*, 9388–9396.
33. Baer, T., Sztáray, B., Kercher, J. P., Lago, A. F., Bodi, A., Skull, C., et al. (2005). *Physical Chemistry Chemical Physics: PCCP, 7*, 1507–1513.
34. Baer, T., Booze, J. A., & Weitzel, K. M. (1991). *Vacuum ultraviolet photoionization and photodissociation of molecules and clusters*. Singapore: World Scientific.
35. Dutuit, O., Baer, T., Metayer, C., & Lemaire, J. (1991). *International Journal of Mass Spectrometry Ion Processes, 110*, 67–82.
36. Jarvis, G. K., Weitzel, K.-M., Malow, M., Baer, T., Song, Y., & Ng, C. Y. (1999). *Review of Scientific Instruments, 70*, 3892–3907.
37. Ng, C. (2000). *Journal of Electron Spectroscopy and Related Phenomena, 112*, 31–46.

Chapter 3
Theory

3.1 Preamble

This chapter is about the computational approaches used in the analysis of the experimental data. Computational methods have been used within the work presented in this thesis in combination with the experimental results, to help untangle the dissociation dynamics, determine thermochemical values and provide a more complete picture of the potential energy surfaces,

(1) Density functional theory (DFT) methods have been used to calculated molecular geometries and potential energy paths.
(2) A few selected different composite methods have been used to determine ion dissociation potentials and thermochemical values.
(3) Using unimolecular rate theory to model the ion breakdown curves produced from the coincidence experiments to determine 0 K onset energies.
(4) To model the TPES using Franck–Condon factors, and elucidate the geometry of the cations.

The GAUSSIAN 03 [1] and GAUSSIAN 09 [2] computational suits have been used throughout the work presented in this thesis. The QChem computational suite [3] was used to calculate ionization levels and excited state potential energy paths shown in Chap. 6 using EOM-IP-CCSD (equation of motion ionization potential coupled cluster singles and doubles) methods [4, 5].

3.2 Computational Methods

3.2.1 Calculating Molecular Geometries and Potential Energy Paths

Optimized geometries were calculated for the neutral parent molecule, parent molecular ion, ionic transition states and the ionic and neutral fragments, using density functional theory (DFT) with the high level of theory and basis set, B3LYP

6-311++G(d,p). B3LYP [6] is the Becke three parameter [7] hybrid Lee–Yang–Parr exchange correlation functional [8], which includes a combination of Hartree–Fock exchange with DFT exchange correlation. 6-311++G(d,p) is a Pople-type basis set, applicable from hydrogen to krypton, with each core atomic orbital basis function comprised of six Gaussian primitive functions. 311 corresponds to three contracted basis functions corresponding to each valence atomic orbital, a triple-ζ basis set. The first one (3) is composed of a linear combination of three primitive Gaussian functions and the second and third (11) are of a linear combination of individual primitive Gaussian functions [9]. Diffuse functions (++) added to each angular momentum function of the basis set which describes the sparse distribution of electrons far from the effects of the nucleus [10]. Diffuse s orbital functions are added to hydrogen, and diffuse s and p orbitals are added to the remaining heavier atoms. The polarization functions (d,p) have higher angular momentum quantum numbers than the occupied atomic orbitals. They are used to describe the distortion of the atomic orbitals within a molecule caused by the interactions with neighbouring atoms, p orbitals are added to the hydrogen, and d orbitals are added to 2nd row atoms [11].

The optimized geometries were used to supply the various input molecular parameters required for modelling the breakdown diagrams and TPES. The optimized ion geometries, which are global minimums or stationary points in the potential energy surface, were used as the starting points for two types of constrained optimizations to generate reaction paths. The first is where a bond may be lengthened to between 3 and 5 Å as for dissociation, and the second where the bond angles can be increased or decreased, as in a rearrangement. The reaction path scans were used to identify possible sites (energies and geometries) along the potential energy surface that are transition state regions, in which the true transition state can be located. Transition states were then accurately located using the TS and Synchronous Transit-guided Quasi-Newton (STQN) methods; QST2 and QST3 integrated into the Gaussian computational suite [12, 13]. The TS method requires an initial guess geometry input for the transition state, found from the constrained optimization scans. For best results, this starting geometry should be fairly close to the one that is required. However, QST2 and QST3 do not require such an accurate initial transition state geometry, but simply the start and end product molecule geometries and in the case of QST3, a less accurate transition state geometry. Once the optimized geometries of the transitions states were found, composite calculations were undertaken to accurately determine their thermochemical properties.

3.2.2 Composite Methods

Thermochemical values such as bond dissociation energies, enthalpies of formation, reaction barrier heights and zero-point vibrational energies can be calculated accurately using a variety of computational methods, all varying in computational

3.2 Computational Methods

cost (time), accuracy and reliability. The first requirement for such calculations is; all calculations must be performed using systems whose geometries have been optimized. That is to say, the point on the potential energy surface at which the first derivative of the energy (the gradient) and hence the negative of this derivative (the forces) is zero. These points, known as stationary points, correspond to any *minimum* (global or local) representing a stable structure, and any *saddle point* or *transition state* (which is a maximum along the reaction coordinate, but a minimum along all other coordinates) present on the potential energy surface, see Fig. 3.1. The most basic of methods consists of a one-step frequency calculation on an optimized stationary point where sources of input for determining thermochemical values are the translational, vibrational, rotational and electronic partition functions. Such input commands typically comprise of which particular basis set and level of theory one wishes to use, in order to calculate the vibrational frequencies e.g. a lower level of theory e.g. the wave functional theory (WFT) Hartree–Fock methods (HF) and basis set such as the minimal STO-3G basis set (one basis function for each atomic orbital with the sum of 3 Gaussian functions approximating a Slater type function) or a higher level of theory (e.g. density functional theory (DFT) with e.g. the B3LYP hybrid functional) and higher basis set with additional polarization diffuse functions added such as the 6–311+G(d,p) basis set. However, the exact exchange–correlation functional, i.e. a unique method to accurately calculate enthalpies of formation or bond dissociation energies using a single determinant alone remains the Holy Grail in computational chemistry [14]. Subsequently modifications have been made, the most successful

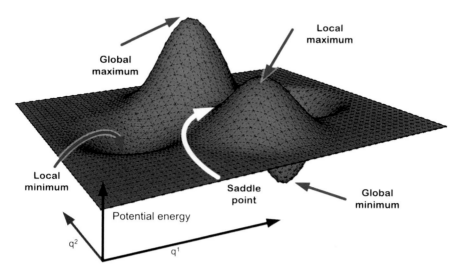

Fig. 3.1 Sketch of a 3-dimensional potential energy surface along reaction coordinates q^1 and q^2. At both minimums and saddle point (transition state) along coordinate q^1, the energy gradients are zero and are termed stationary points

of which are in a class of calculations termed *composite calculations*. Composite methods consist of combining several steps with varying basis sets and levels of theory in an attempt to capitalize the effectiveness of a particular basis set but offset with a lower level of theory. The aim of such composite methods is to reach a compromise between computational cost and reliable, highly accurate results to within ~ 1–10 kJ mol^{-1}. The different types of density functional theory (DFT) extrapolation based composite methods are the Gn methods; G2, G3 [15], G2B3, G3B3 [16], Complete Basis Set (CBS) methods; CBS-QB3, CBS-APNO [17–19], and the hybrid wave function/DFT based Weizmann theory, W1[20] and W2 [21]. Improvement to W1 and W2 is given by the wave function based W3 [22] method. Traditional ab initio quantum chemistry (wave function theory) attempts to improve the reliability and accuracy of determining the molecular energetics with increasing the accuracy of the wave function, at considerable computational cost. An overview of the varying accuracy, cost and treatment of electron correlation across the different methods, from single functional methods like HF and DFT B3LYP to modern ab initio extrapolation methods such as W1-2 [20–22] is shown in Fig. 3.2.

The methods used throughout the work presented in this thesis are the composite method G3B3 and the W1 method. A summary of the main processes involved in each method are presented below.

The G3B3 composite method is an additive method based on a series of six calculations performed under one keyword, G3B3 and is a combination of high level theory and smaller basis sets, with two experimental parameters [15, 16].

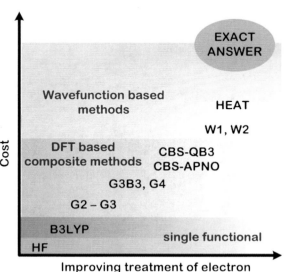

Fig. 3.2 Diagram showing the increasing accuracy across the range of available ab initio methods to accurately calculate thermodynamic properties, towards the exact value

(1) The equilibrium geometry is obtained with density functional theory: B3LYP/6-31G(d). The equilibrium energies are refined using all electrons for the calculation of the correlation energies. The geometry is then used as the basis for the remainder of the calculation. The major difference between the different Gn methods is highlighted here, the geometry is obtained by HF methods for G3, whereas DFT is used in G3B3 and G4.
(2) The harmonic frequencies (scaled by a pre-determined factor) are obtained using the equilibrium geometry, and are used to give the zero point vibrational energy (ZPE).
(3) A group of single point energies are calculated using the following combination of second and fourth order Møller–Plesset and quadratic configuration interaction QCISD(T) levels of theory: MP4(FC)/6-31G(d), MP4(FC)/6-31+G(d), MP4(FC)/6-31G(2df,p), QCISD(T,FC)/6-31G(d),MP2(FU)/G3large. These energies are used for a series of four corrections, (a) a correction is applied for diffuse functions, (b) a correction for higher polarization functions on non-hydrogen atoms and p-functions on hydrogens, (c) correction for correlation effects beyond fourth-order perturbation theory and (d) a correction for the effects of larger basis set and also for non-additivity which is caused by assuming separate basis set extensions for diffuse functions and higher polarization functions. Another difference between G3B3 and other Gn theories, is the combination of what theories are used at this stage; e.g. G3 and G3(MP2) uses mostly Møller–Plesset perturbation theory, MP2, MP4.
(4) The spin–orbit correction, taken from a combination of experiment and theoretical calculations where applicable,[15] is combined with the above four results from step (3) for atomic species only.
(5) A 'higher level correction' is added which accounts for remaining deficiencies within the energy calculation utilizing the corrections for paired and unpaired valence electrons in both atoms and molecules.
(6) Finally, the total energy at 0 K, E_0, is derived by summing the ZPE from step 2, to the result from step 5.

Whilst this method is very accurate for closed shell neutral molecules, it struggles to accurately calculate thermochemical values for molecules containing heavy atoms and open shell species such as cations.

The experimentally derived components of the Gn methods can carry in themselves large uncertainties that are too large to be acceptable for a highly accurate system. Weizmann theory, W1 is based on extrapolation to the complete basis set limit, with separate extrapolations of SCF (self-consistent field, HF), CCSD (coupled cluster with all connected singles and doubles) [23] and (T) (perturbative triple excitation effects) [24] components was developed. W1 has no empirical parameters, empirical additivity corrections and other corrections which are derived, often achieving an accuracy of less than 0.3 kcal mol^{-1} or ca. 1 kJ mol^{-1}. The W1 protocol is outlined as follows;

(1) The geometry is optimized at the B3LYP/VTZ+1 level
(2) Harmonic frequencies are obtained from the geometry optimization, the ZPE calculated and scaled by a factor
(3) Single point calculations using coupled cluster methods; CCSD(T)/AVDZ+2d and CCSD(T)/AVDZ+2d1f
(4) Another single point calculation using CCSD(T)/AVQZ+2d1f is made
(5) The SCF component of total electronic energies is extrapolated
(6) The CCSD valence correlation component is derived
(7) The (T) valence correlation component is derived
(8) The contributions to core correlation are obtained at the CCSD(T) level with the smallest Martin–Taylor basis set (MTvtz, denoted simply as MT).
(9) Finally, scalar relativistic and spin–orbit coupling effects are treated at the ACPF/MT (averaged coupled pair functional) small level [25]. This basis set is the best compromise between quality and computational expense, which is mostly consumed by the core correlation calculations [20, 21].

However, the W1 method is optimized for row 1 and 2 elements and heavier atoms are excluded [26].

Other advancements include the formulation of the high accuracy extrapolated ab initio thermochemistry, HEAT [27], protocol based on the Weizmann theories and large basis sets which is a wave function method devoid of all empirical scaling factors and adjustments [28]. The next step in development is the method as used by Csontos et al. [29], which is based upon a combination of W3 and HEAT. However, these advancements delivering thermochemical accuracies of less than 1 kJ mol^{-1} come at considerable computational cost.

3.3 Modelling

3.3.1 Modelling the Breakdown Curves

The experimental breakdown curves were modelled using the program developed by Sztáray et al. [30]. The program was used to model both slow and fast dissociations. Both the ion time-of-flight distributions and the breakdown curves are modelled using the following fixed parameters; the thermal energy distribution, ionization energy, the parameters which affect the ion TOF i.e. ion acceleration fields and the acceleration and drift distances, and ion vibrational frequencies to determine the density of states of the dissociating ion. The vibrational frequencies of the transition state determine the entropy of activation. Within the Rigid Activated Complex (RAC-)-RRKM framework used throughout this work, selected vibrational modes (the transitional modes) which turn into overall rotational modes of the product ion are scaled by a constant; five frequencies are scaled for the loss of a neutral with 3-D rotations, four for the loss of a linear fragment and two for atom loss [30]. As the low transitional frequencies are scaled, determining their precise values with ab initio calculations is not necessary.

By altering the barrier height and scaling the transition state frequencies, the number and density of states available to the reaction are changed. Altering the temperature changes the width of the ion internal energy and the width of the crossover section of the breakdown curves (where 50 % of parent ions dissociate into fragment daughter ions, starting at the energy range where the parent ion signal initially decreases to where it disappears, the E_0 of a fast reaction), in other words, the slope of the approach to E_0. A higher temperature will broaden the crossover region. A lower temperature will narrow the region, producing a sharper approach to E_0. The actual value of E_0 itself is *independent* of the temperature [31, 32]. The temperature dependence of the breakdown curves means they are effectively a molecular thermometer, measuring the temperature of the neutral molecule [33]. Altering the transition state frequencies effects the density of states; lowering the transition state frequencies increases the reaction rates and increasing them decreases reaction rates.

In some instances the *IE* may also be varied to reproduce the breakdown diagram, as is the case for the fast dissociation of $CFBr_3^+$ [34]. In these instances the ionization potential value supplied from the literature may not be suitable and could have been inaccurately determined from fairly ambiguous photoelectron spectra. As such the *IE* may also be fitted, providing a better fit to the experimental data around the onset energy as well as an improved value for the *IE* [34]. The general process for modelling the breakdown diagram and ion TOF distributions is given in Fig. 3.3.

3.3.1.1 Fast Dissociations

The E_0 for a fast dissociation into the first daughter ion (lowest dissociation channel) is found where the parent ion signal reaches zero. It is assumed that the neutral thermal distribution is faithfully transposed onto the ion manifold, i.e. every parent ion with more internal energy than the dissociation threshold results in a fragment ion. The fractional abundance of the parent ion plotted as a function of photon energy (the breakdown curve) corresponds to the cumulative distribution function (CDF) of the ion internal energy with the dissociation energy as the integration limit, and by inference the CDF of the neutral internal energy at the experimental temperature. The ratio of the parent ion signal is given by;

$$\text{BD}(hv) = \int_0^{E_0-IE} P_i(E, hv)\text{d}E \cong \int_0^{E_0-hv} P_n(E)\text{d}E \quad (3.1)$$

where P_i is the normalized internal energy distribution of the parent ion as a function of the internal and photon energies. P_n is the internal energy of the neutral molecule which can be calculated using the Boltzmann formula; $P_n(E) = \rho_n(E) \cdot e^{-E/kT}$ where ρ_n is the density of states of the neutral molecule,

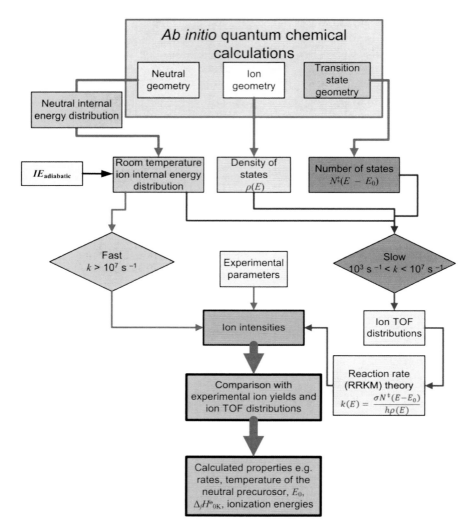

Fig. 3.3 Schematic of the breakdown diagram modelling process. Ab initio calculations provide the input, fast dissociations only require the room temperature ion internal energy distribution whereas slow dissociations also require the ion density of states and the number of states supplied by the transitions state geometry. Ion time-of-flight (TOF) distributions must be fitted to obtain accurate reaction rates and reproduce the breakdown curves. Experimental parameters include the temperature and the dimensions of the experiment

calculated using the molecules vibrational frequencies and rotational constants determined with ab initio calculations. At $h\nu = E_0 - IE$ the above integral becomes zero and the disappearance of the parent ion signal gives the 0 K dissociation onset energy. In other words, E_0 is reached when no part of the thermal distribution is contained within the bound part of the ion potential energy curve any more, but has progressed above the dissociation limit.

3.3 Modelling

The breakdown diagram of a fast dissociation at 0 K should be a step function, where the parent ion signal drops instantly from 100 to 0 % and the daughter ion signal increases from 0 to 100 % at the dissociation limit. In practice, the finite resolution of both the electron and photon spectrometers, together with the molecules thermal distribution of internal energies, means the ion yield curves deviate from the step-function ideal. Instead, the cross-over region can be as large as several hundred meV. However, modelling this breakdown diagram is necessary when the disappearance of the parent ion is poorly defined. Furthermore, modelling the breakdown diagram can supply information about the shape of the potential energy surface by telling us if the neutral thermal distribution is fully accommodated onto the bound part of the ion manifold i.e. into the deep well, before dissociation occurs, or not. If the width of the thermal energy distribution is greater than the depth of the well, then low energy neutrals do not produce ions, and the well is deemed shallow. This is expanded upon in Chap. 4.

The breakdown diagram can be modelled by considering the thermal energy distribution of the neutral molecule, which yields the energy distribution of the ion as a function of photon energy. Frequencies for the ion and neutral were taken from calculations, as outlined in Sect. 3.3.1. The only parameters which can be altered to accurately determine E_0 is the barrier height, the temperature and in a small handful of cases the *IE* value.

3.3.1.2 Slow Dissociations

The window of dissociation rates of iPEPICO, in which the absolute rate constants are measurable, is $10^3 \text{ s}^{-1} < k < 10^7 \text{ s}^{-1}$. Above 10^7 s^{-1}, the dissociation occurs too swiftly to measure the rate, and below 10^3 s^{-1} the dissociation is not accommodated within the time frame of the iPEPICO experiment. As a result of this physical constraint, the slowly dissociating ions do not have enough time to fragment before reaching the detector. Consequently, the ion TOF distribution is not a single narrow Gaussian-type peak, but one that has a diminishing quasi-exponential tail, towards longer TOF times. The whole signal of this metastable ion is attributed to dissociations occurring during the course of acceleration. The signal from metastable daughter ions born in the remainder of the flight tube is so smeared out up to the parent ion TOF limit and is barely perceptible, see Fig. 3.4.

For slow dissociations, the true 0 K onset is not found by inspection of the breakdown diagram where the parent ion signal disappears, but is somewhat lower in energy. A second integral, the ion energy distribution function multiplied by the analytical solution to the differential equation of the unimolecular kinetics, needs to be included giving the equation below [30],

$$\text{BD}_{parent}(hv) = \int_0^{E_0-IE} P_i(E,hv)dE + \int_{E_0-IE}^{+\infty} P_n(E,hv) \cdot \exp(-k(E) \cdot \tau_{max})dE \quad (3.2)$$

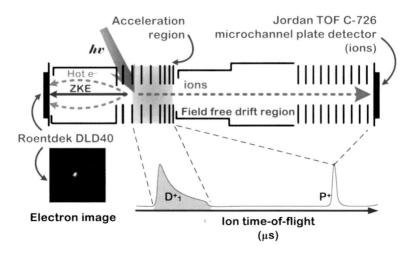

Fig. 3.4 Schematic of the coincidence setup. The metastable daughter D_1^+ is formed very slowly throughout the acceleration region in a quasi-exponential fashion. Any ions including the parent ions, P^+ that are formed at the ionization region produce a symmetrical TOF distribution

here $k(E)$ is the internal energy-dependant rate constant and τ_{max} is the maximum flight time within which the parent ion has to dissociate into the fragment ion in order to be detected as fragment ion [30]. The fragment ion fractional abundance is then given by 1 minus the abundance of the parent ion;

$$BD_{fragment}(h\nu) = \int_{E_0-IE}^{+\infty} P_n(E, h\nu) \cdot (1 - \exp(-k(E) \cdot \tau_{max}))dE \quad (3.3)$$

Modelling the breakdown diagram requires the dissociation rate constants to be taken into account; these can be extracted from the ion TOF distributions. The fragment ion peak shape is given as,

$$Fr_i(h\nu) = \int_{E_0-IE}^{+\infty} P(E, h\nu) \cdot (\exp(-k(E) \cdot \tau(TOF_i)) - \exp(-k(E) \cdot \tau(TOF_{i+1})))dE$$

$$(3.4)$$

The normalized height of the ion peak channel is given by $Fr_i(h\nu)$, $P(E,h\nu)$ is the internal energy distribution of the parent ion, τ is the dissociation time, and $\tau(TOF)$ is the time corresponding to the time of flight channel, i. The sharp TOF spectrum is convoluted with a Gaussian distribution to account for thermal broadening or kinetic energy release, to give the final TOF spectrum [30].

The modelled breakdown curves are determined by using the thermal energy distribution of the molecule and an assumed dissociation rate function, $k(E)$ obtained using statistical RAC-RRKM theory. The adjustable parameters become the barrier height (giving the 0 K onset), the transition state vibrational frequencies and the temperature. A unique $k(E)$ function is obtained from fitting the experimental TOF distributions which is used in conjunction with fitting the other parameters to obtain E_0 [30]. The metastable ion TOF peak distribution depends on the absolute dissociation rate, but the breakdown diagram depends on the ratio of rate constants for parallel dissociations. Therefore it is the fitting of the TOF distributions which is most affected by changing the transition state frequencies, and the shape of the breakdown diagram is less affected.

3.3.1.3 Parallel and Consecutive Reactions

Parallel reactions are those dissociation reactions giving more than one set of product fragments, which are derived from the same parent ion molecule and occurring simultaneously;

$$ABC + h\nu - e^- \rightarrow ABC^+ \rightarrow AB^+ + C \text{ Primary dissociation channel (1)} \qquad (3a)$$

$$ABC + h\nu - e^- \rightarrow ABC^+ \rightarrow A + BC^+ \text{ Parallel dissociation channel (2)} \qquad (3b)$$

The experimental data supplies the relative rates of dissociation which have to be modelled in order to derive the E_0 of the latter parallel reactions (3b). As the photon energy approaches E_0 from above, the rate constant decreases to the limit dictated by the density of states of the ion at the E_0 internal energy. The appearance energy is then given by extrapolation. It is necessary to model the primary dissociation of the parent ion and the resulting breakdown diagram gives the ratio of the dissociation rates in the energy region in which the two processes (the primary and parallel reaction) compete, Eq. (3.5). The relative rates are therefore a function of the number of states of the two transition states;

$$\frac{k_1(E)}{k_2(E)} = \frac{N_1^{\ddagger}(E - E_1)}{N_2^{\ddagger}(E - E_2)} \qquad (3.5)$$

$N_1^{\ddagger}(E - E_1)$ is the sum of the internal energy states in transition state for the reaction 1 from 0 to $E - E_1$. $N_2^{\ddagger}(E - E_2)$ is the corresponding quantity for the parallel reaction. This gives the derived model rate curve for the parallel onset from which the rate is found. As for the primary dissociation, extrapolation to where the rate curve vanishes gives E_0, which is often below the energy at which the increasing signal in the breakdown diagram is observed. Within the rigid-activated-complex (RAC) RRKM framework used to model the parallel onsets in Chap. 5 two parameters are required, (1) the difference of the two E_0 values, and (2) the activation entropy difference (derived from the low vibrational modes of the transition state) to determine the shape, or more specifically the slope of the breakdown curves. If the transition state is loose then the slope has a steeper gradient, conversely if the transition state is tight then the slope is shallower.

Overall, because the shape of the breakdown curve for the parallel reaction is determined by these two parameters, the slope of the onset for the second reaction channel is gentler than the sharper onsets for the primary dissociation reactions. If the reaction is fast, then no experimental information about the absolute rates is supplied. In this instance the vibrational frequencies of one of the transition states is fixed, and the other is varied to fit the relative rates [35]. It must be remembered that as the onset is more gentle, the precision by which the E_0 is determined is reduced, and a greater error must be assigned to it.

Subsequent dissociations of primary daughter ions into further smaller mass ions are termed sequential reactions.

$$ABC + h\nu \rightarrow ABC^+ + e^- \quad \text{Ionization} \quad (3c)$$

$$ABC^+ \rightarrow AB^+ + C \quad \text{Primary dissociation channel} \quad (3d)$$

$$AB^+ \rightarrow A^+ + B \quad \text{Sequential channel} \quad (3e)$$

The treatment of such reactions is different from the above parallel reactions. If the sequential reaction is fast, the yield of the tertiary ion depends upon the energy partitioning in the products of the primary dissociation channel. Similarly, this too can be modelled with statistical theory by assuming that excess energy after dissociation is statistically partitioned amongst the vibrational, rotational and translational degrees of freedom of the primary products AB^+ and C [36]. This means that the breakdown curve of A^+ can be modelled using only the rotational degrees of freedom and vibrational frequencies of the AB^+ and C products, which can be calculated using ab initio methods. The properties of the transition state are ignored [35]. The product ion distribution of A^+ for reaction (3e) formation of $A^+ + B + C$, has a broader internal energy distribution because excess energy is partitioned into both AB^+ and C fragments (reaction 3d) [37]. This manifests itself in the breakdown diagram as a broad and gently curving onset. If the reaction is slow, then fitting both the breakdown diagram and the asymmetric ion TOF distributions of A^+ limits the acceptable range of E_0, meaning it can be determined with a greater precision than the fast parallel dissociations mentioned above [35].

3.3.2 Modelling the TPES

Modelling the TPES using Franck–Condon factors (FCF), the square of the overlap integrals of the ion and neutral wavefunctions, provides the link between ground electronic states of the neutral molecule and its cation. By fitting the TPES, the transition probabilities for ionization into each possible final vibrational state of the ground electronic ion state can be obtained. Noting how the FCFs change with varying cation geometry to provide the best fit to the TPES, can provide us with the best geometry for the cation.

3.3 Modelling

Calculating the FCFs requires the geometries of both the neutral and ion molecules. Highly accurate density functional theory (DFT, B3LYP/6-311++G(d,p)) calculations were performed using the GAUSSIAN 03 suite of programs to obtain the geometry and vibrational frequencies of the ground state neutral, as well as that of the ground state cation. The program 'FCfit v2.8.8' was used to simulate and fit the TPES [38]. A schematic of the modelling procedure is given in Fig. 3.5. The program computes the FC integrals of multidimensional harmonic oscillators based on the recursion formula of Doktorov et al. [39, 40].

The normal coordinates of the ground neutral and ground ion electronic state are related by the linear orthogonal transformation given by Duschinsky, where the normal coordinate of the neutral is given by the normal coordinate of the ion multiplied by the rotation matrix which turns the coordinate system of one state into another, plus the displacement vector [41].

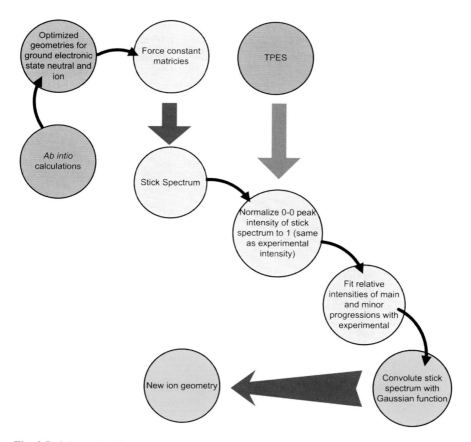

Fig. 3.5 Schematic of the process of modelling the TPES. Ab initio calculations and the experimental data provide the input. Outputs are the convoluted stick spectrum and subsequent new ion geometry

In the first stage, the program FCfit v2.8.8 reads the vibrational coordinates from the optimized geometries for the neutral and cation molecules, together with the ab initio force constant matrices, to calculate a stick spectrum of the possible vibrational progressions based on the normal modes of vibration. Cursory intensities are based upon the initial input neutral and cation geometries, with a vibrational temperature of 0 K. This stick spectrum is used to help assign the peaks in the TPES, as all transitions with non-zero FC intensity are simultaneously given irrespective of their magnitude. As only relative energies of the transitions are determined, the stick spectrum can be aligned to the experimental TPES by the addition of the adiabatic ionization energy. In the second stage, the relative intensities of the vibrational peaks in the major progression (up to a maximum of 10 quanta) are fitted to the experimental intensities by fine tuning the cation geometry; followed by subsequent fitting of the intensities of the weaker progressions. It is accepted generally that ab initio methods are much more robust, accurately describing the potential energy surface for closed shell neutral, than for open shell species. As such, it is the cation geometries which are optimized and the neutral geometry remains unaltered. Finally, the stick spectrum is convoluted with a Gaussian function to simulate the rotational envelope and experimental resolution generating a comparable spectrum to the experimental spectrum. Peaks arising from vibronic coupling (Herzberg–Teller) [42] can be accommodated because the geometries of the ion state are not fixed and are allowed to change. However, as the program uses the properties of harmonic oscillator eigenfunctions, the effects of anharmonicity on the vibrational spacing are not included. Though the program 'FCfit v2.8.8' is to some degree a black box program, some diligence is required to ensure that the correct optimized geometry inputs for both neutral and ion are supplied. The FCF calculation is sensitive to the methods used to calculate the optimized geometries of the neutral and ion states; as such the same method needs to be used for both calculations. In the event of an unexpected geometry change upon ionization, the result can be tested by supplying various optimized cation geometries, and only altering the intensities of the vibrational progressions which do and do not affect the fit. Despite the success of fitting the ground electronic states, fitting the TPES of excited states is more difficult as the geometry of the excited state cation may greatly differ from the neutral geometry. In addition to the difference in geometries, other processes, such as vibronic coupling may be at play, complicating the spectrum.

References

1. Frisch, M. J., Trucks, G. W., Schlegel, H. B., Scuseria, G. E., Rob, M. A., Cheeseman, J. R., et al. (2003). Wallingford, CT: Gaussian, Inc.
2. Frisch, M. J., Trucks, G. W., Schlegel, H. B., Scuseria, G. E., Robb, M. A., Cheeseman, J. R., et al. (2009). *Gaussian 09, Revision A.1.*
3. Shao, Y., Molnar, L. F., Jung, Y., Kussmann, J., Ochsenfeld, C., Brown, S. T., et al. (2006). *Physical Chemistry Chemical Physics, 8*, 3172–3191.

References

4. Kállay, M., & Gauss, J. (2004). *The Journal of Chemical Physics, 121*, 9257.
5. Landau, A., Khistyaev, K., Dolgikh, S., & Krylov, A. I. (2010). *The Journal of Chemical Physics, 132*, 014109–014122.
6. Stephens, P. J., Devlin, F. J., Chabalowski, C. F., & Frisch, M. J. (1994). *Journal of Physical Chemistry, 98*, 11623–11627.
7. Becke, A. D. (1993). *The Journal of Chemical Physics, 98*, 5648–5652.
8. Lee, C., Yang, W., & Parr, R. G. (1988). *Physical Review B, 37*, 758–789.
9. McLean, A. D., & Chandler, G. S. (1980). *The Journal of Chemical Physics, 72*, 5639–5649.
10. Clark, T., Chandrasekhar, J., Spitznagel, G. W., & Schleyer, P. V. R. (1983). *Journal of Computational Chemistry, 4*, 294–301.
11. Frisch, M. J., Pople, J. A., & Binkley, J. S. (1984). *The Journal of Chemical Physics, 80*, 3265–3270.
12. Peng, C., & Schlegel, H. B. (1993). *Israeli Journal of Chemistry, 33*, 449–454.
13. Peng, C., Ayala, P. Y., Schlegel, H. B., & Frisch, M. J. (1995). *Journal of Computational Chemistry, 17*, 49–56.
14. Lazarou, Y. G., Prosmitis, A. V., Papadimitriou, V. C., & Papagiannakopoulos, P. (2001). *Journal of Physical Chemistry A, 105*, 6729–6742.
15. Curtiss, L. A., Raghavachari, K., Redfern, P. C., Rassolov, V., & Pople, J. A. (1998). *The Journal of Chemical Physics, 109*, 7764–7776.
16. Baboul, A. G., Curtiss, L. A., Redfern, P. C., & Raghavachari, K. J. (1999). *The Journal of Chemical Physics, 110*, 7650–7657.
17. Montgomery, J. A, Jr, Ochterski, J. W., & Petersson, G. A. (1994). *The Journal of Chemical Physics, 101*, 5900–5910.
18. Ochterski, J. W., Petersson, G. A., & Montgomery Jr, J. A. (1996). *The Journal of Chemical Physics, 104*, 2598–2620.
19. Montgomery, J. A, Jr, Frisch, M. J., Ochterski, J. W., & Petersson, G. A. (1999). *The Journal of Chemical Physics, 110*, 2822–2828.
20. Martin, J. M. L., & de Oliveira, G. (1999). *The Journal of Chemical Physics, 111*, 1843–1856.
21. Parthiban, S., & Martin, J. M. L. (2001). *The Journal of Chemical Physics, 114*, 6014–6029.
22. Boese, A. D., Oren, M., Atasoylu, O., Martin, J. M. L., Kállay, M., & Gauss, J. (2004). *The Journal of Chemical Physics, 120*, 4129–4142.
23. Purvis, G. D., & Bartlett, R. J. (1982). *The Journal of Chemical Physics, 76*, 1910–1919.
24. Raghavachari, K., Trucks, G. W., Pople, J. A., & Head-Gordon, M. (1989). *Chemical Physics Letters, 157*, 479–483.
25. Gdanitz, R. J., & Ahlrichs, R. (1988). *Chemical Physics Letters, 143*, 413–420.
26. Karton, A., & Martin, J. M. L. (2012). *The Journal of Chemical Physics, 136*, 124114–124126.
27. Harding, M. E., Vázquez, J., Ruscic, B., Wilson, A. K., Gauss, J., & Stanton, J. F. (2008). *The Journal of Chemical Physics, 128*, 114111–114126.
28. Tajti, A., Szalay, P. G., Császár, A. G., Kállay, M., Gauss, J., Valeev, E. F., et al. (2004). *The Journal of Chemical Physics, 121*, 11599–11613.
29. Csontos, J., Rolik, Z., Das, S., & Kállay, M. (2010). *Journal of Physical Chemistry A, 114*, 13093–13103.
30. Sztáray, B., Bodi, A., & Baer, T. (2010). *Journal of Mass Spectrometry, 45*, 1233–1245.
31. Borkar, S., & Sztáray, B. (2010). *Journal of Physical Chemistry A, 114*, 6117.
32. Kercher, J. P., Stevens, W., Gengeliczki, Z., & Baer, T. (2007). *International Journal of Mass Spectrometry, 267*, 159–166.
33. Baer, T., Booze, J. A., & Weitzel, K. M. (1991). *Vacuum ultraviolet photoionization and photodissociation of molecules and clusters*. Singapore: World Scientific.
34. Bodi, A., Kvaran, Á., & Sztáray, B. (2011). *Journal of Physical Chemistry A, 115*, 13443–13451.
35. Baer, T., Sztáray, B., Kercher, J. P., Lago, A. F., Bodi, A., Skull, C., et al. (2005). *Physical Chemistry Chemical Physics: PCCP, 7*, 1507–1513.

36. Baer, T., & Hase, W. L. (1996). *Unimolecular reaction dynamics: Theory and experiments.* New York: Oxford University Press Inc.
37. Sztáray, B. L., & Baer, T. (2002). *Journal of Physical Chemistry A, 106,* 8046–8053.
38. Spangenberg, D., Imhof, P., & Kleinermanns, K. (2003). *Physical Chemistry Chemical Physics: PCCP, 5,* 2505–2514.
39. Doktorov, E. V., Malkin, I. A., & Man'ko, V. I. (1975). *Journal of Molecular Spectrometry, 56,* 1–20.
40. Doktorov, E. V., Malkin, I. A., & Man'ko, V. I. (1977). *Journal of Molecular Spectrometry, 64,* 302–326.
41. Duschinsky, F. (1937). *Acta Physicochimica U.R.S.S., 7,* 551–556.
42. Herzberg, G., & Teller, E. Z. (1933). *Physical Chemistry B, 21,* 410.

Chapter 4
Fast Dissociations of Halogenated Methanes: A Thermochemical Network

4.1 Preamble

The work presented in this chapter has been published as a journal article entitled 'A Halomethane Thermochemical Network from iPEPICO Experiments and Quantum Chemical Calculations' in 2012 by J. Harvey, R. P. Tuckett and A. Bodi, in the Journal of Physical Chemistry A, volume 116, issue 39, pages 9696–9705. The majority of the data collection and analysis was performed by the author, however, the assistance lent by Ms Nicola Rogers, Drs Mathew Simpson Andras Bodi, Melanie Johnson and Professor Richard Tuckett during beamtime with the collection of the data is gratefully acknowledged. The modelling program was developed by Sztáray et al. [1]. The threshold photoelectron spectra can be found in Appendix C.

4.2 Introduction

This chapter focuses on the study of fast dissociations of halogenated methanes using threshold photoion photoelectron coincidence techniques. It will be demonstrated how the primary piece of information yielded from such experiments, the 0 K onset energy E_0 for the production of the first photodissociation daughter ion, can be used to construct a network comprised of enthalpies of formation of neutral and ion species from which more updated and new thermochemical values can be derived.

Eleven enthalpies of formation are updated, including that for $CBrClF_2$. Importantly, enthalpies of formation reported in the literature for several neutral and ion species which were derived using purely computational methods are also confirmed.

Calculations were performed using the GAUSSIAN 09 computational suit [2]. Rate constants calculated with Rice–Ramsperger–Kassel–Marcus (RRKM) theory at arbitrarily chosen transition state geometries along the dissociation coordinate show that the dissociations were confirmed to be fast (rates in excess of 10^7 s^{-1}). Fast dissociative photoionization processes in threshold PEPICO experiments are modelled simply by taking into consideration the thermal energy distribution of the neutral molecule, which yields the energy distribution of the ion as a function of photon energy [1].

G3B3 [3] and W1 [4] composite methods (see Chap. 3) were used to determine the neutral and ion energetics, which were utilised along with previously reported energies [5] in the construction of the thermochemical network shown in Sect. 4.3.4. The experimental onset energies in the thermochemical network provide rigid links between the neutral and the ion enthalpies of formation. In the shallow well instances, the initial abundance of the first daughter ion is non-zero. Even in such cases, the fit was required to reproduce the disappearance energy range of the parent signal, thus giving the E_0 value. The photon energy, at which the deep well approximation fails, then yields the adiabatic IE.

Quantum chemical calculations on small molecules can yield thermochemical values with a few kJ mol^{-1} uncertainties or better, often outperforming experimental results [5–9]. Recent threshold photoelectron photoion coincidence (TPEPICO) experiments, in which both the photon and the photoelectron energies are known to within 1–2 meV (0.1–0.2 kJ mol^{-1}), [1, 10–12] are capable of measuring dissociative photoionization onset energies in small to medium sized molecules with such levels of accuracy. To reiterate, in the absence of an overall reverse barrier, the onset energies, E_0, correspond to the reaction energy at 0 K, and yield the enthalpies of formation for the parent ion, daughter ion or neutral fragment if two out of the three are known;

$$E_0 = \Delta_f H_{0K}[ion] + \Delta_f H_{0K}[neutral\ fragment] - \Delta_f H_{0K}[neutral\ parent\ molecule] \tag{4.1}$$

Recent advances in ab initio methods can be rigorously tested and confirmed by results derived from experiment. The two approaches are, thus, complementary and can be applied simultaneously to provide sturdier results. For example, experiment can be used to confirm the highly accurate enthalpies of formation derived by Csontos et al. [5], for a comprehensive range of neutral halogenated methanes, which were inspired by the W3 and HEAT protocols [13, 14].

In threshold photoelectron photoion coincidence, unimolecular dissociation reactions of internal energy selected parent ions are studied as a function of photon energy, yielding daughter ion appearance energies [15]. Ions are mass analysed in delayed coincidence with threshold electrons, and the breakdown diagram is generated by plotting the fractional abundance of parent and fragment ions as a function of $h\nu$. For a fast dissociation, every parent ion with more internal energy than the dissociation threshold results in a fragment ion, and the breakdown curve of the parent ion corresponds to the cumulative distribution function (CDF) of the ion internal energy to the dissociation energy. In the first approximation, it follows that the breakdown curve corresponds to the CDF of the internal energy of the neutral at the experimental temperature:

$$\mathrm{BD}(h\nu) = \int_0^{E_0-IE} P_i(E, h\nu)\mathrm{d}E \cong \int_0^{E_0-h\nu} P_n(E)\mathrm{d}E \qquad (4.2)$$

where P_i is the normalised internal energy distribution of the parent ion as a function of the internal and photon energies. P_n is the internal energy distribution of the neutral molecule, calculated using the Boltzmann formula $P_n(E) = \rho_n(E) \cdot \mathrm{e}^{-E/kT}$, where $\rho_n(E)$ is the density of states of the neutral.

The above integral vanishes at $h\nu = E_0$. Consequently, the 0 K appearance energy, E_0, is given by the disappearance energy of the parent ion in small molecules, and modelling the breakdown diagram only requires the internal energy distribution of the neutral precursor [1]. Two assumptions are made when modelling the breakdown curve. First, the neutral internal energy distribution is transposed directly onto the ion manifold; in other words the threshold ionization cross sections for the sequence transitions are constant over the thermal energy range, and there is a uniform probability of threshold ionization across the neutral molecule's energy distribution [1, 12, 16]. The Franck–Condon factors for sequence transitions in small molecules at room temperature are dominated by rotational contributions and this assumption will hold true as long as the geometries of the neutral and parent ion are sufficiently similar. However, it is important to note that no assumption is made about the Franck–Condon factors for threshold ionization as a function of photon energy. Second, the second integral in Eq. (4.2) assumes that zero internal energy neutrals always contribute to the parent ion signal. This is only valid if the ground state potential well is deep enough to accommodate the transposition of the entire thermal energy distribution of the neutral onto the ion manifold in the photon energy range of the breakdown diagram, which will be termed the 'deep well assumption' (Fig. 4.1). However, if the width of the thermal energy distribution is larger than the depth of the potential energy well, the low energy neutrals do not contribute to the ion signal. Now, the 'deep well assumption' is no longer valid and what we term a 'shallow well reality' prevails. In such cases, the parent signal is always less than 100 %, and there is significant daughter ion signal, *even* at the ionization limit. This effect had been observed previously [17], and was first discussed in the TPEPICO study of $CFBr_3$ and CBr_4 [11].

While a reasonable estimate for the 0 K onset can be trivially deduced for fast dissociations of small molecules, modelling the breakdown curve provides a more rigorous assessment of the assumptions, confidently confirming the shape and nature of the ion internal energy distribution and the validity of the deep well assumption. Even though the latter is applicable in most covalently bound ions, it is not always appropriate in weakly bound systems, a few of which we will examine in this work. In such cases, the adiabatic ionization energy (IE) can often be derived from the breakdown diagram.

As can be seen from Eq. (4.2), the modelled breakdown curves are temperature dependent and the breakdown diagram is effectively a molecular thermometer, measuring the temperature of the neutral molecule. Additionally, oscillations or

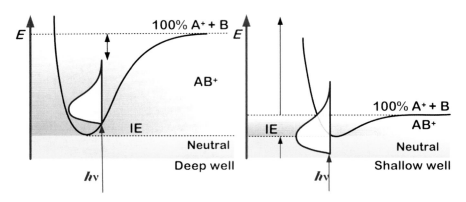

Fig. 4.1 Schematic showing the deep well and the shallow well scenarios. As the photon energy ($h\nu$) is scanned, the parent ion (AB^+) fractional abundance corresponds to the normalized parent ion internal energy distribution integral from the bottom of the well to the barrier ($A^+ + B$). The daughter ion (A^+) fractional abundance is given as the portion of the thermal distribution from the dissociation limit to infinity, and in the shallow well scenario this integral is always non-negligible. For photon energies below the adiabatic IE, the breakdown diagram deviates from the shape predicted by the deep well assumption as the neutral thermal distribution cannot be fully incorporated onto the ionic manifold, so not all neutral molecules produce parent ions

peaks in the breakdown curve, as was first observed for CH_3I^+, [10] may indicate changes in the threshold photoionization mechanism.

We employ calculations, together with experimental 0 K dissociative photoionization onsets, to derive a self-consistent thermochemical network, which links neutral and ionic species and provides improved enthalpies of formation, and re-affirms the results of previous theoretical and experimental studies [5, 18]. Ion thermochemical values can be useful in the interpretation of ion dynamics, e.g. Selected Ion Flow Tube experiments [19]. Such self-consistent networks, not dissimilar to the Active Thermochemical Tables of Ruscic et al. [20], add to the expanding thermochemical armamentarium available to the researcher. The advantage of this approach over the purely ab initio route is that the network is pegged by the accurately measured onset energies and well-known enthalpies of formation such as those for CF_4 and CH_4, thus reducing the plasticity of the ab initio network and eliminating systematic errors inherent in the original ab initio calculations. Such an approach has also been used to obtain updated enthalpies of formation for primary amines [21], as well as for bromofluoromethanes and their dissociative photoionization products [11].

The dissociative photoionization of $CBrClF_2$ has, to the best of our knowledge, not been reported before, and its enthalpy of formation is not well known. We also seek to provide a more complete thermochemistry for the fragment ion CHF_2^+ by the dissociative photoionization of CH_2F_2 and $CHClF_2$, the latter of which has been studied before at inferior photon resolution using TPEPICO by Howle et al. [22].

4.3 Results and Discussion

The experimental results on the halogenated methanes are presented and discussed first. The dissociative photoionization reactions of each neutral studied are listed through 4a–g

$$CH_3Cl + h\nu \rightarrow CH_2Cl^+ + H + e^- \quad (4a)$$

$$CH_2Cl_2 + h\nu \rightarrow CH_2Cl^+ + Cl + e^- \quad (4b)$$

$$CHCl_3 + h\nu \rightarrow CHCl_2^+ + Cl + e^- \quad (4c)$$

$$CH_3F + h\nu \rightarrow CH_2F^+ + H + e^- \quad (4d)$$

$$CH_2F_2 + h\nu \rightarrow CHF_2^+ + H + e^- \quad (4e)$$

$$CHClF_2 + h\nu \rightarrow CHF_2^+ + Cl + e^- \quad (4f)$$

$$CBrClF_2 + h\nu \rightarrow CClF_2^+ + Br + e^- \quad (4g)$$

As the potential wells for the ground electronic states of $CHCl_3^+$, $CHClF_2^+$ and $CBrClF_2^+$ are quite shallow, it is possible to derive adiabatic ionization energies from modelling the breakdown diagram. Some general remarks about the preferential loss of atoms from the halogenated methanes are now made. The lowest energy dissociative photoionization channel observed in the TPEPICO experiment corresponds to the cleavage of the weakest bond in the parent ion. Thus, with $CBrClF_2$ it is the C–Br bond that breaks first in dissociative photoionization to give $CClF_2^+$; in fully halogenated chlorofluoromethanes CCl_nF_{4-n} ($n = 1-3$), it is always a C–Cl bond that breaks first. When the halomethane molecule incorporates hydrogen atoms, the situation is not so clear cut. For CH_3F and CH_2F_2, H-loss and cleavage of a C–H bond occurs at the lowest energy. Similarly, the C–H bond is the weakest and breaks at the lowest energy in CH_3Cl, whereas in the other two chloromethanes, CH_2Cl_2 and $CHCl_3$, a C–Cl bond breaks first. By contrast, the electronic ground state of the parent ion is repulsive in the Franck–Condon region in CCl_4, CF_4 and CHF_3 [23–25]. It is tempting to try and draw conclusions from these observations, especially with regard to the parent ion potential well depths (Fig. 4.2). How do the well depths vary with increasing homogenous halogenation and inhomogeneous halogenation? Only very general remarks can be made. Firstly, that CF_4, CHF_3, $CFCl_3$ and CCl_4 which are not included in this study, dissociatively photoionize forming no initial parent ion, and one can regard the ion potential well depths as so shallow they do not exist. In another study on CF_nBr_{4-n}, Bodi et al. [11]. found that CBr_4^+ *is* formed and dissociated into $CBr_3^+ + Br$. However the ion potential well is shallow, like the ions $CHCl_3^+$, $CHClF_2^+$ and $CBrClF_2^+$. Based on the cursory observation that the parent ion potential well is deepest for those molecules with more hydrogen atoms, onset energies from threshold coincidence experiments for a range of halogenated methanes can found in the literature, in addition to those included here [11, 15, 26–29]. CH_3Br^+ has the deepest potential well and CH_4^+ is the next most stable ion. With the exception of CH_3Br^+, the parent ions are then destabilised with each substitution of an exisiting hydrogen atom with a halogen atom. The parent ions are further stabilized with each successive substitution of an existing halogen for another. In the absence of any hydrogen atoms, bromine atom substitution has the most stabilizing effect. Further experiment would be required to obtain the full set of data, and to assess the relative well depths and stability of the parent ion the 0 K onsets for secondary daughter ion formation would also need to be considered.

4.3.1 Chlorinated Methanes

The breakdown diagram of CH_3Cl in the 12.88–13.05 eV photon energy range is shown in Fig. 4.3a. The fitted E_0 for $CH_3Cl \rightarrow CH_2Cl^+ + H + e^-$, i.e. the 0 K appearance energy of the first daughter ion is 12.981 ± 0.004 eV. Tang et al. [30] also observed CH_2Cl^+ as the first daughter ion at these low energies, which is in contrast with photoionization by He(I) radiation at 21.2 eV in which no CH_2Cl^+ was detected below 15 eV [24, 31]. Autoionization is only possible in tuneable VUV studies, such as this work. Because the H-loss channel opens up in a Franck–Condon gap, it is not accessible by direct photoionization, e.g. by using He(I) radiation, at threshold. On the other hand, the hydrogen 'many-line' light source in an earlier tuneable non-threshold photoionization study by Werner et al. [32] may have provided insufficient flux around 13 eV for the H-loss channel to be observed.

The breakdown diagram of CH_2Cl_2 in the 11.85–12.20 eV photon energy range is shown in Fig. 4.3b. The derived E_0 for reaction (4b) is 12.108 ± 0.003 eV. Our value is more accurate and somewhat lower than the 12.122 ± 0.010 eV reported by Lago et al. [26] obtained using a larger photon energy step size in the onset region. We note that both reactions 3a and 3b share a common fragment ion, CH_2Cl^+.

The breakdown diagram of $CHCl_3$ is given in Fig. 4.3c across the 11.20–11.60 eV photon energy range. The fitted E_0 is 11.487 ± 0.005 eV. The $CHCl_2^+$ daughter ion is produced in reaction (4c), and its abundance at the literature IE of 11.3 eV [33] is non-zero meaning that a significant proportion of parent ions have sufficient energy to dissociate, as was also reported by Shuman et al. [34]. If the Franck–Condon factors for threshold ionization are assumed to be uniform across the thermal energy distribution, the breakdown curve may be modelled as the ratio of the integrated internal energy distribution *in* the potential well against that *above* the well. Because the portion of the distribution which falls below the IE does not contribute to the ion signal, the IE also influences the breakdown curve. Alternatively, instead of retro-fitting the IE to reproduce the breakdown curve, a deep potential well is assumed and the point at which the calculated curve deviates from the experimental curve can be taken as

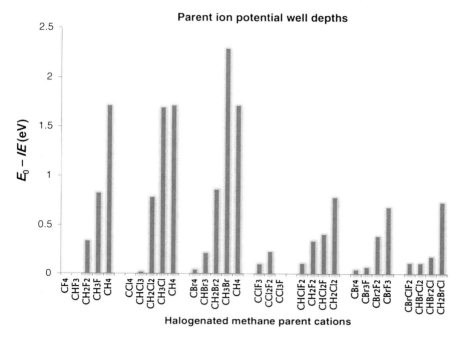

Fig. 4.2 Bar chart of the depths of the potential wells (E_0–IE) for parent ions. Onset and adiabatic ionization energies taken from literature sources and from this work

the adiabatic IE of the parent molecule. This was also the approach used in a study of $CFBr_3$ and CBr_4 [11], as well as for $C_2H_4I_2$ [17], where the point of deviation from the modelled curve led to a revised IE for these three molecules. In such cases, only initially hot neutrals are ionized at $h\nu < IE$.

Neutrals with less internal energy than ($IE-h\nu$) are not ionized, and the parent ion internal energy does not correspond to the Boltzmann distribution of the neutral in this energy range [17]. The advantage of using this modelling method to determine the IE in such instances, as opposed to modelling the TPES by Franck–Condon simulations [35] or ab initio calculations, is its considerable ease of use. Furthermore, thanks to the enhanced resolution of the experiment the deviation of the calculated fit and the experimental points at 11.47 ± 0.01 eV can clearly be seen, providing a new value for the IE of $CHCl_3$. As Fig. 4.3c shows, the adiabatic IE is not easily identifiable from the threshold photoelectron spectrum (TPES) of the molecule, as the ground state of $CHCl_3^+$ is not vibrationally resolved, unlike that of CH_3Cl and CH_2Cl_2 [33]. The signal onset for the ground state of the ion does appear to be 11.3 eV, which agrees with an 11.30 ± 0.05 eV appearance energy given by Seccombe et al. [36]. from a lower-resolution TPEPICO study; they also quoted a vertical ionization energy of 11.51 eV. Based on this latest breakdown diagram, we propose that the rise of the threshold electron signal below

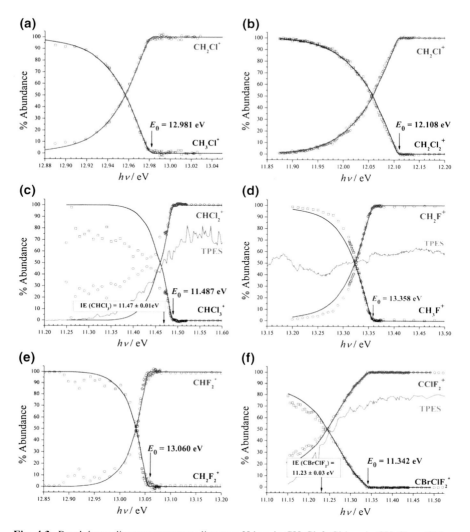

Fig. 4.3 Breakdown diagram corresponding to **a** H loss in CH_3Cl, **b** Cl loss in CH_2Cl_2, **c** Cl loss in $CHCl_3$, **d** H loss in CH_3F, **e** H loss in CH_2F_2 and **f** Br loss in $CBrClF_2$. The experimental points (*open shapes*) are plotted together with the modelled breakdown curves (*solid line*). The derived 0 K onset energies are shown together with the new ionization energies for shallow well parent ions $CHCl_3^+$ and $CBrClF_2^+$

11.47 eV is due to hot bands. Please see Appendix B for all input parameters for modelling the breakdown diagrams.

4.3.2 Fluorinated Methanes

Figure 4.3d shows the breakdown diagram of CH_3F in the 13.20–13.50 eV photon range. Two aspects are noted. First, the 0 K onset coincides with a small but sharp rise in the TPES signal. As was previously reported [10], a proposed explanation for such rises seen in the pulsed field ionization (PFI-)PEPICO experiments of methane [37] due to Rydberg state lifetime considerations, are unlikely to be at play in continuous field experiments. Consequently, this step function is probably due to an additional threshold photoionization channel which opens up at the onset and enhances daughter ion production. Second, and as a consequence of the first point, the whole breakdown curve cannot be faithfully modelled using a single temperature and the corresponding neutral internal energy distribution. The lower energy range, below 13.325 eV, is reproduced when 298 K is assumed while the parent ion abundances above the crossover are overestimated. A higher temperature of 348 K models the higher energy points well but widens the breakdown diagram too much. However, it is the latter temperature which leads to a perfect fit in the most important energy region, namely at the disappearance energy of the parent signal. The fitted E_0 for reaction 4d is 13.358 ± 0.005 eV, much higher than the room temperature appearance energy (the photon energy at which daughter ion signal above the noise is first observed) quoted by Weitzel and coworkers as 13.20 ± 0.08 eV [38]. From their observed IE from the TPES of 12.53 eV and the 13.34 ± 0.02 eV crossover energy, they derived the dissociation energy of CH_3F^+ to be 0.84 ± 0.02 eV, and an E_0 of 13.37 ± 0.02 eV. They also reported a non-vanishing parent ion signal, i.e. one that does not decrease to 0 % above the 0 K appearance energy. Our own crossover and onset energies are both ca. 20 meV lower. This discrepancy could result from insufficient hot electron suppression in the Weitzel study.

The breakdown diagram of CH_2F_2 in the 12.85–13.20 eV photon energy range is presented in Fig. 4.3e. The fitted E_0 of reaction 3e is 13.060 ± 0.015 eV, with the larger error limit being a result of the curve having a small gradient near the E_0. This produces a tailing off of the parent signal, instead of a sharp cut-off. A possible reason for this and the slightly inferior signal-to-noise ratio at the onset energy compared with other molecules in this study may lie with the subtraction of the hot electron contamination. If the threshold electron yield is low at onset or changes quickly with photon energy, the ring area around the detector (which is subtracted from the centre area signal) can be a poor representation of the hot electron background in the centre. While the E_0 is independent of sample temperature, the shape of the curve at the onset is governed by it, and a softly-landing parent ion curve leads to a less well defined E_0. Using pulsed-field-ionization zero kinetic energy electrons (PFI-ZEKE), Forysinski et al. [39] found the 0 K H-loss appearance energy to be 13.065 ± 0.003 eV. This value from their laser-based very high resolution study is 5 meV higher than our own reported value, but the two values are within the error limits. Both values are somewhat lower than the appearance energy of CHF_2^+ at 298 K of 13.08 ± 0.03 eV reported by Seccombe et al. [40] from a lower resolution TPEPICO study.

4.3.3 CBrClF$_2$ and CHClF$_2$

The breakdown diagram of $CBrClF_2$ in the 11.15–11.50 eV photon energy range along with the TPES is presented in Fig. 4.3f. The C–Br bond is the weakest, and therefore the first dissociative photoionization channel produces $CClF_2^+$ + Br + e^-. The fitted E_0 value is 11.342 ± 0.003 eV. There is some ambiguity as to the adiabatic IE for $CBrClF_2$. The TPES has been studied several times, and vertical ionization energies of 11.51[41] and 11.83 eV [42] were reported. Another value for the IE is reported to be 11.21 eV, corresponding to the onset of the electron signal [43]. However, as can be seen from the TPES, the identification of the adiabatic IE is not immediately obvious. Similarly to $CHCl_3$ and $CHClF_2$, there is a significant daughter ion contribution of ca. 30 % at the ionization onset. The modelled breakdown curve deviates from the experimental data points at a slightly higher energy of 11.23 ± 0.03 eV. We propose this somewhat higher value as the adiabatic IE of $CBrClF_2$. This fit is less sensitive to the assumed IE than for $CHCl_3$ and $CHClF_2$, as the deep well model only deviates at most over a few tens of meV from the shallow well reality.

The breakdown diagram of $CHClF_2$ (Fig. 4.4), is shown in the 12.15–12.51 eV photon energy range, plotted together with the corresponding TPES. The C–Cl bond is the weakest, and the derived E_0 value for reaction 4 g, production of CHF_2^+, is 12.406 ± 0.004 eV. This is the same daughter ion as the first product from CH_2F_2. As with $CHCl_3$, there is significant daughter ion abundance present in the low energy region of the breakdown diagram, and the deep well approximation deviates from the experimental points over the range of 12.2–12.3 eV. This provides us with an accurate IE of 12.30 ± 0.02 eV, which is not easily determined from the TPES alone. The most recent ionization energies range from 12.15 ± 0.05 eV determined by TPES, [22] to 12.28 ± 0.02 eV determined by PIMS [43], so our value is at the higher end of this range. Upon magnification, a small hump or bulge in the otherwise smooth breakdown curve becomes apparent at 12.37 eV. Interestingly, this feature is also faithfully reproduced by the model curves. To shed light on this peculiarity, we also plotted the calculated neutral thermal energy distribution in Fig. 4.4, with its origin shifted to 12.37 eV. Two possible reasons have been suggested for such features in breakdown diagrams, namely alternative photoionization mechanisms [10] or a less than faithful transposition of the neutral thermal energy distribution to the ion manifold [1]. Here, however, the explanation appears to be that the thermal energy distribution is indeed transposed onto the ion manifold, and we use our molecular thermometer to measure the Boltzmann distribution of neutral energies, which shows a dip at an internal energy of ca. 50 meV due to the higher density of rovibrational levels 15 meV higher.

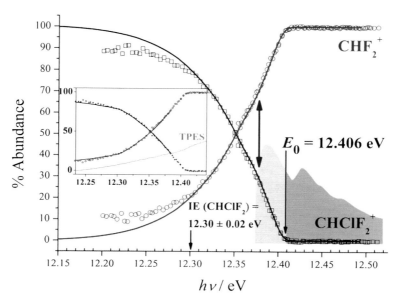

Fig. 4.4 Breakdown diagram corresponding to Cl loss in the shallow well parent ion CHClF$_2^+$, along with the new IE of 12.30 ± 0.02 eV. The ionization energy is hard to predict based on the broad featureless TPES shown in grey (*inset*). The *inset* also shows breakdown diagram modelled with the new IE, where the lower $h\nu$ range is reproduced well by taking the actual potential energy well depth into account. The ion internal energy distribution at 12.37 eV is shown with parent contribution in *grey* and CClF$_2^+$ contribution in *pale blue*. The hump in the breakdown curve, indicated by a *double-headed arrow* at 12.37 eV, corresponds to an unusual minimum in the thermal energy distribution in the neutral, confirming the faithful transposition of internal energy distribution upon threshold photoionization

4.3.4 Thermochemistry

In addition to the measured 0 K appearance energies, the neutral parents and daughter ions can be related to each other through a series of quantum chemical calculations involving closed shell species, as well, to generate a network shown in Fig. 4.5. The neutral network is composed of three sub-networks, that of the chlorinated methanes, the fluorinated methanes and the chlorofluoromethanes. Even though the accurate onset energy of CCl$_4$, CF$_4$ and CHF$_3$ cannot be determined experimentally because they photoionize dissociatively, their neutral enthalpies of formation feature in the network through computed reaction energies. Two stand-alone compounds, CHClF$_2$ and CBrClF$_2$, are also studied, which only connect to the rest of the network via their E_0 values and as well as a neutral isodesmic reaction energy for the former. Each sub-network contains the independent isodesmic reactions connecting the neutrals, e.g. CH$_n$X$_{4-n}$, as well as the ions, e.g. CH$_n$X$_{3-n}^+$, (X = F, Cl),

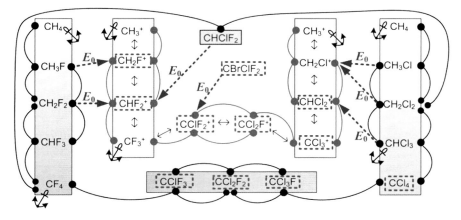

Fig. 4.5 The thermochemical network showing the experimental (E_0) and computational links. Enthalpies of formation are indicated as nodes. The appearance energies (*straight dashed arrows*) connect the neutrals with the ions, ab initio isodesmic (*pairs of curved links*), $H_2/(F_2$ or $Cl_2)$ and F_2/Cl_2 exchange reactions (*double headed arrows*) connect neutral/neutral, and ion/ion groups. Revised thermochemical values are indicated by *dashed boxes*. The absolute values of $\Delta_f H^0_{0K}$ are tethered to five anchor points: CH_4, CF_4, $CHCl_3$, CH_3^+ and CF_3^+

$$CH_n X_{4-n} + CH_{n-2} X_{6-n} \rightarrow 2\, CH_{n-1} X_{5-n} (n = 2-4) \quad (4h)$$

$$CH_n X_{3-n}^+ + CH_{n-2} X_{5-n}^+ \rightarrow 2\, CH_{n-1} X_{4-n}^+ (n = 2, 3) \quad (4i)$$

and the interconnecting exchange reactions for the ionic fragments where a new halomethane is generated by the substitution of a hydrogen with a halogen (e.g. H_2/X_2),

$$CH_n X_{3-n}^+ + \tfrac{1}{2} X_2 \rightarrow CH_{n-1} X_{4-n}^+ + \tfrac{1}{2} H_2 (n = 1-3) \quad (4j)$$

This approach was also adopted for the chlorofluoromethanes: $CCl_n F_{4-n}$ and $CCl_{n-1} F_{4-n}^+$. The enthalpies of formation derived in this work are given in Table 4.1.

In isodesmic reactions, the products and reactants have *equal* number of the same *types* of bonds [49, 50]. It has been shown that these reactions yield reliable reaction energies, especially for closed shell species [50, 51]. Quantum chemical calculations such as the composite calculations used in this chapter often suffer from error accumulation arising from; basis set truncation, inaccuracies calculating valence electron correlation and zero-point vibrational energies [52], and in heavier atoms, difficulty with accurate recovery of spin–orbit coupling effects [6, 9]. These errors as well as relativistic and core correlation effects are effectively cancelled out in isodesmic reaction energy calculations [49, 50].

The network is essentially a graph of vertices (enthalpies of formation of ions and neutrals) connected by edges. There are three types of edges; E_0 onset

4.3 Results and Discussion

Table 4.1 Derived and literature values for enthalpies of formation and thermal enthalpies, in units of kJ mol^{-1}

Species	$\Delta_f H^\theta_{ok}$	$\Delta_f H^\theta_{ok}$	$\Delta_f H^\theta_{298k}$†	$H_{298k} - H_{ok}$
CH_4		-66.56 ± 0.06^b	-74.55 ± 0.06^b	10.0
CH_3F		-228.5 ± 2.0^c	-236.9 ± 2.0^c	10.1^c
CH_2F_2		-442.6 ± 2.0^c	-450.5 ± 2.0^c	10.6^c
CHF_3		-687.7 ± 2.0^c	-694.9 ± 2.0^c	11.5^c
CF_4		-927.8 ± 1.3^c	-933.8 ± 1.3^c	12.8
CH_3^+		1099.35 ± 0.1^d	1095.60 ± 0.1^d	10.0
CH_2F^+	844.4 ± 2.1^a	837.0^e	840.4 ± 2.1^a	10.0
CHF_2^+	601.6 ± 2.7^a		598.4 ± 2.7^a	11.0
CF_3^+		413.4 ± 2.0^f	410.2 ± 2.0^f	11.1
$CClF_3$	-702.1 ± 3.5^a	-703.4 ± 3.1^c	-707.3 ± 3.5^a	13.7^c
CCl_2F_2	-487.8 ± 3.4^a	-487.9 ± 4.2^c	-492.1 ± 3.4^a	14.8^c
CCl_3F	-285.2 ± 3.2^a	-282.7 ± 5.3^c	-288.6 ± 3.2^a	15.9^c
$CClF_2^+$	552.2 ± 3.4^a		549.5 ± 3.4^a	11.8
CCl_2F^+	701.2 ± 3.3^a		699.0 ± 3.3^a	12.5
CH_3Cl		-74.3 ± 3.1^c	-82.6 ± 3.1^c	10.4^c
CH_2Cl_2		-88.66 ± 1.3^g	-95.7 ± 1.3^g	11.8^c
$CHCl_3$		-98.4 ± 1.1^h	-103.4 ± 1.1^h	14.1^c
CCl_4	-94.0 ± 3.2^a		-96.4 ± 3.2^a	17.1^c
CH_2Cl^+		961.1 ± 1.7^i	957.1 ± 1.7^i	10.1
$CHCl_2^+$	890.3 ± 2.2^a	891.7 ± 1.5^j	887.2 ± 2.2^a	11.3
CCl_3^+	849.8 ± 3.2^a	852.3 ± 2.5^k	848.3 ± 3.2^a	13.3
$CBrClF_2$	-446.6 ± 2.7^a	-423.8 ± 15^l	-457.6 ± 2.7^a	15.7^m
$CHClF_2$		-475.7 ± 3.1^c	-482.2 ± 3.1^c	12.3
Cl		119.6^g	121.3^g	6.28
F		77.3^g	79.4^g	6.5

† Conversion to 298 K is made using Chase NIST-JANAF compendium values for thermal enthalpies [44], ($H_{298K}-H_{0K}$[C] = 1.05, $H_{298K}-H_{0K}$[H$_2$] = 8.47, $H_{298K}-H_{0K}$[F$_2$] = 8.82, $H_{298K}-H_{0K}$[Cl$_2$] = 9.18 and $H_{298K}-H_{0K}$[Br$_2$] = 24.5 kJ mol^{-1}). 298 K values for cations are obtained using the ion convention, $H_{298K}-H_{0K}$[e$^-$] = 0 kJ mol^{-1}
[a] This work, [b] Ruscic active thermochemical tables [20], [c] Csontos et al. [5] result confirmed by W1 calculation, [d] Bodi et al. [10], [e] Lias et al. [45], [f] Bodi et al. [11] [g] Chase NIST-JANAF compendium [44], [h] Manion [46], [i] Lago et al. [26], [j] Shuman et al. [29], [k] Hudgens et al. [47], [l] Burcat and Ruscic [48], [m] G3B3 value

energies, neutral and ion isodesmic reaction energies, and exchange reaction energies (the latter only for fragment ions). The E_0 onset energies interconnect the neutral and ion groups, whereas the isodesmic and exchange reaction energies establish connections within the neutral and ion groups. If available, isodesmic reaction energies were obtained based on the energy values reported by Csontos et al. [5]. Otherwise, W1 calculated reaction energies were used. G3B3 results were also obtained and checked for consistency, see Table 4.2 (please also see Appendix A). As a starting point in the optimization of the enthalpies of formation, the isodesmic, exchange and dissociative photoionization reaction energies were also calculated using the literature $\Delta_f H^\theta_{0K}$ values.

The entire network needs to be tethered to an absolute scale of enthalpies of formation, and well-defined enthalpies of formation for a chosen set of species provide this vital link. These 'anchor' enthalpies of formation are kept unchanged during the fitting process: $\Delta_f H^{\theta}_{0K}$ (CH$_4$) = -66.56 ± 0.06 [53], $\Delta_f H^{\theta}_{0K}$ (CF$_4$) = -927.8 ± 1.3 [5, 53], $\Delta_f H^{\theta}_{0K}$ (CF$_3^+$) = 413.4 ± 2.0 [11], and $\Delta_f H^{\theta}_{0K}$ (CHCl$_3$) = -98.4 ± 1.1 kJ mol^{-1} [46]. The enthalpy of formation for CH$_3^+$ is derived from the 0 K onset energy for CH$_4 + h\nu \rightarrow$ CH$_3^+ +$ H $+$ e$^-$, 14.323 ± 0.001 eV [27] combined with the H atom heat of formation, which yield $\Delta_f H^{\theta}_{0K}$(CH$_3^+$) = 1099.35 ± 0.1 kJ mol^{-1}, yet another anchor value. The two major groups of neutrals, the fluorinated methanes and chlorinated methanes are 'anchored' by CF$_4$/CH$_4$ and CHCl$_3$/CH$_4$, respectively. The chlorofluoromethane groups, CCl$_n$F$_{4-n}$ and CCl$_{n-1}$F$^+_{4-n}$, provide bridges between the chlorinated and the fluorinated methanes.

An overall error function, (ε) is defined as the sum of component isodesmic, exchange and experimental errors:

$$\varepsilon = \sum_i \varepsilon_i(\text{iso}) + \sum_j \varepsilon_j(\text{exc}) + \sum_k \varepsilon_k(\text{iPEPICO}) \qquad (4.3)$$

where $\varepsilon_i(\text{iso}) = \left(\Delta_r H^{\circ}[\text{calc}]_{\text{iso}} - \Delta_r H^{\circ}\left[\Delta_f H^{\circ}\right]_{\text{iso}}\right)^2, \varepsilon_j(\text{exc}) = \left(\Delta_r H^{\circ}[\text{calc}]_{\text{exc}} - \Delta_r H^{\circ}\left[\Delta_f H^{\circ}\right]_{\text{exc}}\right)^2$ and $\varepsilon_k(\text{iPEPICO}) = \left(\Delta_r H^{\circ}[\text{meas}]_{\text{iPEPICO}} - \Delta_r H^{\circ}\left[\Delta_f H^{\circ}\right]_{\text{dissoc. photoionization}}\right)^2$. $\Delta_r H^{\circ}[\text{calc}]$ are the reaction enthalpies at 0 K from *ab intio* calculations, and $\Delta_r H^{\theta}\left[\Delta_f H^{\theta}\right]$ are the enthalpy of formation based reaction enthalpies, the starting values for which are taken from the literature. The network is optimized by minimizing this error function using the Generalized Reduced Gradient Method [54].

Initially, the network is relaxed and all enthalpies of formation except the anchor values are set as fit parameters. The exchange reaction error function is then weighted by 0.01 to make sure it influences the fit less than the isodesmic reaction errors, as the latter are more reliable. There are fewer experimental onset energies, which are also more accurate than the calculations. Therefore, the PEPICO error function is weighted by 100 to ensure its adequate representation in the optimization of the network, (see Tables 1.2, 1.3, 1.4 and 1.5 of Appendix A). Thus, no data set over-influences the outcomes. Next, by analysing the optimized enthalpies of formation in the fully relaxed fit, we identified further literature values which could be kept constant in the fitting procedure. Notably, the enthalpies of formation reported by Csontos et al. [5]. were compared with the optimized values. The average difference was found to be 0.05 kJ mol^{-1}, with a standard deviation of 0.77 kJ mol^{-1} and maximum of 1.35 kJ mol^{-1} for CH$_3$Cl, CH$_3$F, CH$_2$F$_2$, CHF$_3$, and CHClF$_2$. These enthalpies of formation are, thus, confirmed as recommended by Csontos et al. [5] and held constant in the final fit. Furthermore, the Chase value for $\Delta_f H^{\theta}_{0K}$ (CH$_2$Cl$_2$) = -88.66 ± 1.3 kJ mol^{-1} and the Lago value for $\Delta_f H^{\theta}_{0K}$ (CH$_2$Cl$^+$) = -961.1 ± 1.7 kJ mol^{-1} are kept unchanged [26, 44]. The remaining enthalpies of formation for CH$_2$F$^+$, CHF$_2^+$, CClF$_2^+$,

4.3 Results and Discussion

Table 4.2 Enthalpies of reaction used for the thermochemical network in this chapter

Reaction	(a) G3B3 0 K $\Delta_r H^\circ$ (eV)	(b) CBS-QB3 0 K $\Delta_r H^\circ$ (eV)	(c) W1 0 K $\Delta_r H^\circ$ (eV)	(d) E_0 (eV)	(e) CSONTOS (eV)	(f) W1 values or Csontos values where avaliable $\Delta_r H^\circ$ (eV)	(g) From [litetature $\Delta_f H^\circ$] $\Delta_r H^\circ$ eV	(h) ERROR (f)−(g)
First onsets								
$CHCl_3 \to CHCl_2^+ + Cl$	11.497	11.545	11.542	11.487		11.487	11.495	9.70E−08
$CH_2Cl_2 \to CH_2Cl^+ + Cl$	12.098	12.130	12.163	12.108		12.108	12.120	1.34E−02
$CH_3Cl \to CH_2Cl^+ + H$	12.935	12.962	13.020	12.981		12.981	12.970	1.17E−04
Isodesmic reactions								
$CCl_3^+ + CH_2Cl^+ \to 2CHCl_2^+$	−0.291	−0.291	−0.312			−0.312	−0.139	6.88E−07
$CHCl_2^+ + CH_3^+ \to 2\,CH_2Cl^+$	−0.724	−0.707	−0.701			−0.701	−0.707	2.09E−06
$CCl_4 + CH_2Cl_2 \to 2CHCl_3$	−0.093	−0.099	−0.136		−0.146	−0.141	−0.200	8.52E−23
$CHCl_3 + CH_3Cl \to 2CH_2Cl_2$	−0.020	−0.023	−0.055		−0.042	−0.049	−0.052	3.26E−05
$CH_2Cl_2 + CH_4 \to 2CH_3Cl$	0.069	0.067	0.034			0.034	0.071	1.18E−03
Exchange reactions								
$CCl_3^+ + 1/2H_2 \to CHCl_2^+ + 1/2Cl_2$	−0.546	−0.563	−0.604			−0.604	−0.414	6.18E−06
$CHCl_2^+ + 1/2H_2 \to CH_2Cl^+ + 1/2Cl_2$	−0.255	−0.272	−0.292			−0.292	−0.274	6.60E−06
$CH_2Cl^+ + 1/2H_2 \to CH_3^+ + 1/2Cl_2$	0.469	0.435	0.409			0.409	0.433	5.87E−06

(continued)

Table 4.2 (continued)

Reaction	(a) G3B3 0 K $\Delta_rH°$(eV)	(b) CBS-QB3 0 K $\Delta_rH°$(eV)	(c) W1 0 K $\Delta_rH°$(eV)	(d) E_0(eV)	(e) CSONTOS (eV)	(f) W1 values or Csontos values where avaliable$\Delta_rH°$ (eV)	(g) From [litetature $\Delta_fH°$] $\Delta_rH°$ eV	(h) ERROR (f)-(g)
First onsets								
$CH_3F \to CH_2F^+ + H$	13.364	13.370	13.370	13.358		13.358	13.277	2.53E−05
$CH_2F_2 \to CHF_2^+ + H$	13.085	13.062	13.077	13.055		13.055	13.217	1.29E−04
Isodesmic reactions								
$CF_3^+ + CH_2F^+ \to 2CHF_2^+$	−0.555	−0.555	−0.559			−0.559	−0.173	5.43E−05
$CHF_2^+ + CH_3^+ \to 2CH_2F^+$	−0.107	−0.105	−0.098			−0.098	−0.445	8.24E−04
$CF_4 + CH_2F_2 \to 2CHF_3$	−0.088	−0.078	−0.089		−0.053	−0.053	−0.052	8.36E−07
$CHF_3 + CH_3F \to 2CH_2F_2$	0.310	0.328	0.318		0.322	0.322	0.322	4.10E−07
$CH_2F_2 + CH_4 \to 2CH_3F$	0.558	0.576	0.561		n/a	0.561	0.541	4.21E−04
Exchange reactions								
$CF_3^+ + 1/2H_2 \to CHF_2^+ + 1/2F_2$	0.550	0.560	0.560			0.560	0.512	2.27E−05
$CHF_2^+ + 1/2H_2 \to CH_2F^+ + 1/2F_2$	1.106	1.115	1.119			1.119	1.078	1.62E−05
$CH_2F^+ + 1/2H_2 \to CH_3^+ + 1/2F_2$	1.212	1.220	1.216			1.216	1.205	1.34E−06

(continued)

4.3 Results and Discussion

Table 4.2 (continued)

Reaction	(a) G3B3 0 K $\Delta_r H°$ (eV)	(b) CBS-QB3 0 K $\Delta_r H°$ (eV)	(c) W1 0 K $\Delta_r H°$ (eV)	(d) E_0 (eV)	(e) CSONTOS (eV)	(f) W1 values or Csontos values where avaliable $\Delta_r H°$ (eV)	(g) From [litetature $\Delta_f H°$] $\Delta_r H°$ eV	(h) ERROR (f)-(g)
First onsets								
$CF_3Cl \rightarrow CF_3^+ + Cl$	12.861	n/a	12.823	12.801		12.801	12.801	1.58E − 05
$CF_2Cl_2 \rightarrow CClF_2^+ + Cl$	12.079	n/a	12.068	12.03		12.03	12.01	7.96E − 02
Isodesmic reactions								
$CF_3^+ + CCl_2F^+ \rightarrow 2CClF_2^+$	−0.124		−0.107			−0.107	−0.107	5.76E − 17
$CCl_3^+ + CClF_2^+ \rightarrow 2CCl_2F^+$	−0.003		0.005			0.005	0.005	4.16E − 17
$CF_4 + CCl_2F_2 \rightarrow 2CF_3Cl$	0.091		0.091		0.091	0.091	0.118	7.00E − 04
$CClF_3 + CFCl_3 \rightarrow 2\,CF_2Cl_2$	0.103		0.103		0.108	0.108	0.121	1.75E − 04
$CF_4 + CHCl_3 \rightarrow CCl_3F + CHF_3$	0.457				0.539	0.539	0.552	1.75E − 04
Exchange reactions								
$CF_3^+ + 1/2Cl_2 \rightarrow CF_2Cl^+ + 1/2F_2$	0.983		1.058			1.058	0.999	3.38E − 05
$CF_2Cl^+ + 1/2Cl_2 \rightarrow CFCl_2^+ + 1/2F_2$	1.107		1.164			1.164	1.106	3.38E − 05
$CFCl_2^+ + 1/2Cl_2 \rightarrow CCl_3^+ + 1/2F_2$	1.110		1.160			1.160	1.102	3.38E − 05

(continued)

Table 4.2 (continued)

Reaction	(a) G3B3 0 K $\Delta_r H°$(eV)	(b) CBS-QB3 0 K $\Delta_r H°$(eV)	(c) W1 0 K $\Delta_r H°$(eV)	(d) E_0(eV)	(e) CSONTOS (eV)	(f) W1 values or Csontos values where avaliable $\Delta_r H°$ (eV)	(g) From [litetature $\Delta_f H°$] $\Delta_r H°$ eV	(h) ERROR (f)-(g)
First onsets								
$CHClF_2 \to CHF_2^+ + Cl$	12.448		12.435	12.406	12.406	12.406	12.404	1.63E−04
$CHCl_2F \to CHFCl^+ + Cl$	12.945		13.019	11.909	Not included			
Isodesmic reactions								
$CF_4 + CH_2Cl_2 \to 2CHClF_2$	0.622		0.644		0.652	0.652	0.674	4.89E−04
First onsets								
$CBrClF_2 \to CClF_2^+ + Br$	11.401			11.342	11.342	11.342	11.342	2.98E−16

Column (h) is included as an example of the difference between enthalpies of reaction determined using the average of the calculated enthalpies of reaction (f) and enthalpies of reaction determined using literature values of the enthalpies of formation of the constituent molecules. Column (h) varies with each iteration of the global fit to minimize the sum of errors (sum of column (h))

4.3 Results and Discussion

CCl_2F^+, $CHCl_2^+$, CCl_3^+, CF_3Cl, CF_2Cl_2, $CFCl_3$, $CBrClF_2$ and CCl_4 are the final fit parameters. The network is anchored to CF_4 on the left hand side in Fig. 4.5, but, since CCl_4 is a fit parameter, an alternative, asymmetric anchor is required for the chloromethane series. As the Manion value for $\Delta_f H^0_{0K}$ ($CHCl_3$) [46], -98.4 ± 1.1 kJ mol^{-1}, has been used previously as an anchor value by Shuman et al. [29] and has a lower error bar than the Csontos value of -94.6 ± 5.3 kJ mol^{-1} we have also opted for it to act as anchor. Having established the anchor values, confirmed the most reliable literature enthalpies of formation which are also kept unchanged, and set the error function weights to construct a balanced fit, a final optimization was carried out to obtain the final results as summarized in Table 4.1.

Three sources determine the uncertainties of the final results: (1) the uncertainties in the anchor values which peg the network to the enthalpy of formation scale; (2) the calculation errors; and (3) the errors in the iPEPICO appearance energies. These were accounted for as follows: (1) the anchor values were set to the high and low limit of their confidence interval, and a relaxed fit was carried out establishing the network confidence interval for each optimized species; (2) a ± 2 kJ mol^{-1} uncertainty contribution was assumed for calculations; (3) a further ± 2 kJ mol^{-1} uncertainty contribution was assumed for species, on which we have no direct experimental appearance energy data; (4) the iPEPICO appearance energy uncertainty was used otherwise. The confidence intervals listed in Table 4.1 are the result of these four contributions.

The conversion from 0 K to 298 K is made by the relationship [44]:

$$\Delta_f H_{298K} = \Delta_f H_{0K} + [H_{298K} - H_{0K}]_{molecule} - [H_{298K} - H_{0K}]_{constituent\ elements} \quad (4.4)$$

The list of thermal enthalpies, $H_{298K}-H_{0K}$, is shown in Table 4.1. The W1 values are virtually identical to those from the Csontos et al. [5] study. The latter are used for the neutral molecules when available, but the G3B3 value is used for $CBrClF_2$. W1 values are used for the remaining neutrals and the fragment cations.

First, we consider data for the chlorinated and fluorinated methanes. The 0 K enthalpy of formation of CCl_3^+ is connected to the enthalpies of formation of CH_2Cl^+ and $CHCl_2^+$ via isodesmic and exchange pathways. Using well-established values for $\Delta_f H^0_{0K}(CHCl_3) = -98.4 \pm 1.1$ kJ mol^{-1}, [46] and $\Delta_f H^0_{0K}(CH_2Cl_2) = -88.7 \pm 1.3$ kJ mol^{-1} [44], our global fit derives revised values for $\Delta_f H^0_{0K}(CHCl_2^+) = 890.3 \pm 2.2$ and $\Delta_f H^0_{0K}(CCl_3^+) = 849.8 \pm 3.2$ kJ mol^{-1}. The $CHCl_2^+$ value is similar to that determined by Shuman et al. [29] 891.7 ± 1.5 kJ mol^{-1}. The revised $\Delta_f H^0_{0K}$ (CCl_3^+) value is about 15 kJ mol^{-1} higher than the 834.6 kJ mol^{-1} quoted by Lias [45], and an earlier value of Rodriguez et al. [55] of 831.6 kJ mol^{-1}. However it agrees with the Robles et al. [56] value of 847.68 ± 3.3 kJ mol^{-1} derived from $\Delta_f H^0_{0K}(CCl_3) = 69.8 \pm 2.5$ kJ mol^{-1} and their measured IE(CCl_3) = 8.06 ± 0.02 eV. It is also within the error limit of the Hudgens et al. [47] value of $\Delta_f H^0_{0K}(CCl_3^+) = 852.3 \pm 2.5$ kJ mol^{-1}, also derived from photoionization of CCl_3.

CH_2F^+ and CHF_2^+ are fitted parameters. They connect in the network to the fixed values of $\Delta_f H_{0K}^\theta$ of CH_3F, CH_2F_2 and $CHClF_2$ via their experimental 0 K appearance energies. We note that an experimental value for the enthalpy of formation of CH_3F has been surprisingly hard to determine, with Chase et al. [44] quoting a value of -226 ± 33 kJ mol^{-1}. One year earlier, Luo and Benson [57] had recommended the 'best' experimental value at 298 K to be -233.9 ± 4.2 kJ mol^{-1}, corresponding to -225.5 ± 4.2 kJ mol^{-1} at 0 K Given the importance and relative simplicity of this five-atom halogenated hydrocarbon, it is perhaps surprising that the range of theoretical values in the literature is also large. Values at 0 K from -224 to -230 kJ mol^{-1} have been reported by many authors, with errors spanning from ± 0.8 to ± 10.0 kJ mol^{-1} [5, 18, 48, 58–60]. As explained above, we have fixed the 0 K value for CH_3F to that determined by Csontos et al. [5] -228.5 ± 2.0 kJ mol^{-1}. By contrast, the Chase and Csontos values for CH_2F_2 are almost equal, with similar errors of ca. ± 2.0 kJ mol^{-1}. The heat of formation of $CHClF_2$ is reported by Csontos et al. [5] to be -475.7 ± 3.1 kJ mol^{-1}, with no obvious experimental value for comparison. The new value for $\Delta_f H_{0K}^\theta$ (CH_2F^+) of 844.4 ± 2.1 kJ mol^{-1} is significantly higher than that reported by Lias et al. [45] 837.0 kJ mol^{-1}. The new value for $\Delta_f H_{0K}^\theta$ (CHF_2^+), 601.6 ± 2.7 kJ mol^{-1}, is equivalent to 598.4 ± 2.7 kJ mol^{-1} at 298 K. This latter value is in reasonable agreement with a recent experimental value at 298 K from Seccombe et al. [40, 61] of 604 ± 3 kJ mol^{-1}, where the appearance energy at this temperature of CHF_2^+ from CH_2F_2 was corrected for thermal effects by the procedure of Traeger et al. [62].

The experimental route is important here because reliable quantum chemical calculations involving bromine containing compounds are difficult to perform [59] due to the large numbers of electrons, and the fact that relativistic effects become significant for high–Z atoms [18, 63, 64]. Although Borkar et al. [65] have found that relative energies of C_3H_5Br isomers can be quite well predicted by standard computational methods, bromine and iodine containing species are omitted in the comprehensive study of Csontos et al. [5], and Bodi et al. [11] report large error bars, typically in the region of 7 kJ mol^{-1} in a recent study on bromofluoromthanes. The study of $CBrClF_2$ provides a link between bromine containing species and the remainder of the lighter Cl and F containing species. Little is known about the enthalpy of formation of $CBrClF_2$, indeed the only value we could find was -423.8 ± 15 kJ mol^{-1} given by Burcat [48]. As such, the heat of formation becomes a 'fit' parameter, which is only connected to the network by the E_0 of the reaction $CBrClF_2 \rightarrow CClF_2^+ + Br + e^-$. Barring an overall reverse barrier, the result, $\Delta_f H_{0K}^\theta$ ($CBrClF_2$) = -446.6 ± 2.7 kJ mol^{-1}, falls just outside the generous error limit of the previous value.

The enthalpy of formation of CCl_4 given by Csontos et al. [5] of -88.7 ± 6.4 kJ mol^{-1} lies toward the less negative end of the literature values and has the largest error limit among the values they derived. We derive a revised more negative value for $\Delta_f H_{0K}^\theta$ of -94.0 ± 3.2 kJ mol^{-1}. This value is in excellent agreement to the Rodgers et al. [66]. value of -93.7 ± 0.6 kJ mol^{-1} and the

Chase value of -93.8 ± 2.1 kJ mol^{-1} [44]. We note that Csontos et al. [5] seem to report more reliable enthalpies of formation for fluorine substituted methanes such as CHF$_3$, than when methane is substituted with multiple chlorine atoms such as in CHCl$_3$ and CCl$_4$.

Due to the lack of certainty regarding the enthalpies of formation of CCl$_3$F, CCl$_2$F$_2$ and CClF$_3$ in the literature, these values were also fitted. The resulting enthalpies are; $\Delta_f H^0_{0K}$(CCl$_3$F) $= -285.2 \pm 3.2$, $\Delta_f H^0_{0K}$(CCl$_2$F$_2$) $= -487.8 \pm 3.4$ and $\Delta_f H^0_{0K}$ (CClF$_3$) $= -702.1 \pm 3.5$ kJ mol^{-1}. These values are within the uncertainty limits of the Csontos et al. [5] values, namely, -282.7 ± 5.3, -487.9 ± 4.2 and -703.4 ± 3.1 kJ mol^{-1}, respectively. The Chase values of $\Delta_f H^0_{0K}$(CCl$_3$F) $= -285.5 \pm 6.3$ kJ mol^{-1} and $\Delta_f H^0_{0K}$(CClF$_3$) $= -702.8 \pm 3.3$ kJ mol^{-1} are also in close agreement with our results [44]. Whilst there are no experimental results leading directly to CCl$_2$F$^+$ or CClF$_2^+$ in this work, their 0 K enthalpies of formation have been determined by isodesmic and exchange reaction energies to be 701.2 ± 3.3 and 552.2 ± 3.4 kJ mol^{-1}, respectively.

4.4 Conclusions

The thermochemistry of the halogenated methanes CH$_3$Cl, CH$_2$Cl$_2$, CHCl$_3$, CH$_3$F, CH$_2$F$_2$, CHClF$_2$ and CBrClF$_2$, and their fragment ions CH$_2$Cl$^+$, CHCl$_2^+$, CCl$_3^+$, CH$_2$F$^+$, CHF$_2^+$, CCl$_2$F$^+$ and CClF$_2^+$ was studied using a combination of experimental data from imaging photoelectron photoion coincidence spectroscopy and ab initio calculations of isodesmic and exchange reaction energies. A thermochemical network was constructed, in which the neutral and ionic components were intra-connected by sub-networks of isodesmic and exchange reactions, and interconnected by the experimental 0 K dissociative photoionization energies. The network was anchored by the well-known enthalpies of formation for CF$_4$, CH$_4$, CHCl$_3$, CF$_3^+$ and CH$_3^+$. An error function was defined between measured dissociative photoionization onsets and calculated reaction energies on the one hand, and the reaction energies derived using the enthalpies of formation in the network vertices on the other. The optimum values for the enthalpies of formation were determined by minimizing this error function. This holistic approach has been successful in producing updated thermochemical values at 0 K for the neutrals CCl$_4$, CBrClF$_2$, CClF$_3$, CCl$_2$F$_2$ and CCl$_3$F, as -94.0 ± 3.2, -446.6 ± 2.7, -702.1 ± 3.5, -487.8 ± 3.4 and -285.2 ± 3.2 kJ mol^{-1}, respectively. Fitting the remaining neutral enthalpies of formation led to negligible changes. These selected values were held constant, and are therefore confirmed. Revised 0 K enthalpies of formation for the ions CH$_2$F$^+$, CHF$_2^+$, CClF$_2^+$, CCl$_2$F$^+$, CHCl$_2^+$ and CCl$_3^+$ have been determined as 844.4 ± 2.1, 601.6 ± 2.7, 552.2 ± 3.4, 701.2 ± 3.3, 890.3 ± 2.2 and 849.8 ± 3.2 kJ mol^{-1}, respectively. The adiabatic IEs can easily be obtained based on the breakdown diagram of weakly bound parent ions that only exist in a Franck–Condon allowed shallow potential energy well. These have been found to be 11.47 ± 0.01 eV, 12.30 ± 0.02 eV and 11.23 ± 0.03 eV for $CHCl_3$, $CHClF_2$ and $CBrClF_2$, respectively. We suggest that this is the experimental method of choice to determine the IE of molecules where the ground state of the parent ion is only weakly bound. Finally, because of an uncharacteristic dip in the density of states, the thermal energy distribution of $CHClF_2$ shows a minimum at 50 meV. This interesting feature is also seen and modelled as a small hump at 12.37 eV in the otherwise smooth breakdown curve for $CHClF_2^+$.

References

1. Sztáray, B., Bodi, A., & Baer, T. (2010). *Journal of Mass Spectrometry, 45*, 1233–1245.
2. Gaussian 09. (2009). Revision A.1, Gaussian, Inc.: Wallingford CT.
3. Baboul, A. G., Curtiss, L. A., Redfern, P. C., & Raghavachari, K. J. (1999). *Journal of Chemical Physics, 110*, 7650–7657.
4. Martin, J. M. L., & de Oliveira, G. (1999). *Journal of Chemical Physics, 111*, 1843–1856.
5. Csontos, J., Rolik, Z., Das, S., & Kállay, M. (2010). *Journal of Physical Chemistry A, 114*, 13093.
6. Peterson, K. A., Feller, D., & Dixon, D. A. (2012). *Theoretical Chemistry Accounts, 131*, 1079–1099.
7. Karton, A., Daon, S., & Martin, J. M. L. (2011). *Chemical Physics Letters, 510*, 165–178.
8. Feller, D., Peterson, K. A., & Dixon, D. A. (2011). *Journal of Physical Chemistry A, 115*, 1440–1451.
9. Feller, D., Peterson, K. A., & Grant Hill, J. (2011). *Journal of Chemical Physics, 135*, 044102–044120.
10. Bodi, A., Shuman, N. S., & Baer, T. (2009). *Physical Chemistry Chemical Physics: PCCP, 11*, 11013–11021.
11. Bodi, A., Kvaran, Á., & Sztáray, B. (2011). *Journal of Physical Chemistry A, 115*, 13443–13451.
12. Borkar, S., & Sztáray, B. (2010). *Journal of Physical Chemistry A, 114*, 6117–6123.
13. Boese, A. D., Oren, M., Atasoylu, O., Martin, J. M. L., Kállay, M., & Gauss, J. (2004). *Journal of Chemical Physics, 120*, 4129–4142.
14. Tajti, A., Szalay, P. G., Császár, A. G., Kállay, M., Gauss, J., Valeev, E. F., et al. (2004). *Journal of Chemical Physics, 121*, 11599–11613.
15. Baer, T., Sztáray, B., Kercher, J. P., Lago, A. F., Bodi, A., Skull, C., et al. (2005). *Physical Chemistry Chemical Physics: PCCP, 7*, 1507–1513.
16. Kercher, J. P., Stevens, W., Gengeliczki, Z., & Baer, T. (2007). *International Journal of Mass Spectrometry, 267*, 159–166.
17. Baer, T., Walker, S. H., Shuman, N. S., & Bodi, A. (2012). *Journal of Physical Chemistry A, 116*, 2833–2844.
18. Lazarou, Y. G., Papadimitriou, V. C., Prosmitis, A. V., & Papagiannakopoulos, P. (2002). *Journal of Physical Chemistry A, 106*, 11502–11517.
19. Simpson, M., & Tuckett, R. P. (2012). *Journal of Physical Chemistry A, 116*, 8119–8129.
20. Ruscic, B. (2012). Active Thermochemical Tables, early beta 1.110. Retrieved May 09, 2012, from http://atct.anl.gov/index.html.
21. Bodi, A., Kercher, J. P., Bond, C., Meteesatien, P., Sztáray, B., & Baer, T. (2006). *Journal of Physical Chemistry A, 110*, 13425–13433.
22. Howle, C. R., Collins, D. J., Tuckett, R. P., & Malins, A. E. R. (2005). *Physical Chemistry Chemical Physics: PCCP, 7*, 2287–2297.

23. Smith, D. M., Tuckett, R. P., Yoxall, K. R., Codling, K., & Hatherly, P. A. (1993). *Chemical Physics, 216*, 493–502.
24. Simm, I. G., Danby, C. J., Eland, J. H. D., & Mansell, P. I. (1976). *Journal of Chemical Society, Faraday Transactions 2, 72*, 426–434.
25. Parkes, M. A., Chim, R. Y. L., Mayhew, C. A., Mikhailov, V. A., & Tuckett, R. P. (2006). *Molecular Physics, 104*, 263–272.
26. Lago, A. F., Kercher, J. P., Bodi, A., Sztáray, B., Miller, B., Wurzelmann, D., et al. (2005). *Journal of Physical Chemistry A, 109*, 1802–1809.
27. Weitzel, K.-M., Malow, M., Jarvis, G. K., Baer, T., Song, Y., & Ng, C. Y. (1999). *Journal of Chemical Physics, 111*, 8267–8270.
28. Song, Y., Qian, X.-M., Lau, K.-C., Ng, C. Y., Liu, J., & Chen, W. (2001). *Journal of Chemical Physics, 115*, 4095–4104.
29. Shuman, N., Zhao, L. Y., Boles, M., Baer, T., & Sztáray, B. (2008). *Journal of Physical Chemistry A, 112*, 10533–10538.
30. Tang, X., Zhou, X., Wu, M., Liu, S., Liu, F., Shan, X., et al. (2012). *Journal of Chemical Physics, 136*, 034304–034312.
31. Lane, I. C., & Powis, I. (1993). *Journal of Physical Chemistry, 97*, 5803–5808.
32. Werner, A. S., Tsai, B. P., & Baer, T. (1974). *Journal of Chemical Physics, 60*, 3650–3657.
33. Von Niessen, W., Åsbrink, L., & Bieri, G. (1982). *Journal of Electron Spectroscopy Related Phenomena, 26*, 173–201.
34. Schuman, N., Zhao, L. Y., Boles, M., Baer, T., & Sztáray, B. (2008). *Journal of Physical Chemistry A, 112*, 10533–10538.
35. Innocenti, F., Eypper, M., Lee, E. P. F., Stranges, S., Mok, D. K. W., Chau, F.-T., et al. (2008). *Chemistry–A European Journal, 14*, 11452–11460.
36. Seccombe, D. P., Chim, R. Y. L., Jarvis, G. K., & Tuckett, R. P. (2000). *Physical Chemistry Chemical Physics: PCCP, 2*, 769–780.
37. Weitzel, K.-M., Jarvis, G. K., Malow, M., Baer, T., Song, Y., & Ng, C. Y. (2001). *Physical Review Letters, 86*, 3526–3529.
38. Weitzel, K.-M., Güthe, F., Mähnert, J., Locht, R., & Baumgärtel, H. (1995). *Chemical Physics, 201*, 287–298.
39. Forysinski, P. W., Zielke, P., Luckhaus, D., & Signorell, R. (2010). *Physical Chemistry Chemical Physics: PCCP, 12*, 3121–3130.
40. Seccombe, D. P., Tuckett, R. P., & Fisher, B. O. (2001). *Journal of Chemical Physics, 114*, 4074–4088.
41. Cvitaš, T., Klasinc, L., & Novak, I. (1980). *International Journal of Quantum Chemistry, 18*, 305–313.
42. Douct, J., Gilbert, R., Sauvageau, P., & Sandorfy, C. (1975). *Journal of Chemical Physics, 62*, 366–369.
43. Lias, S. G., Bartmess, J. E., Liebman, J. F., Holmes, J. L., Levin, R. D., & Mallard, W. G. (2011). Ion Energetics Data. In P. J. Linstrom & W. G. Mallard (Eds.), *NIST Chemistry WebBook, NIST Standard Reference Database Number 69*; (p. 20899). Gaithersburg, MD: National Institute of Standards and Technology.
44. Chase, M. W. (1988). *Journal of Physical Chemistry Reference Data., Monograph 9*.
45. Lias, S. G., Bartmess, J. E., Liebman, J. F., Holmes, J. L., Levin, R. D., & Mallard, W. G. (1988). *Journal of Physical Chemistry Reference Data., 17*.
46. Manion, J. A. (2002). *Journal of Physical Chemistry Reference Data, 31*, 124–165.
47. Hudgens, J. W., Johnson, R. D. I., Timonen, R. S., Seetula, J. A., & Gutman, D. (1991). *Journal of Physical Chemistry, 95*, 4400–4405.
48. Burcat, A., & Ruscic, B. (September 2005). Third Millennium Ideal Gas and Condensed Phase Thermochemical Database for Combustion with Updates from Active Thermochemical Tables, ANL-05/20 and TAE 960, Technion-IIT; Aerospace Engineering, and Argonne National Laboratory, Chemistry Division. Retrieved September 2011, from ftp://technion.ac.il/pub/supported/aetdd/thermodynamics.

49. Wheeler, S. E., Houk, V., Schleyer, P. V. R., & Allen, W. D. (2009). *Journal of the American Chemical Society, 131*, 2547–2560.
50. Wodrich, M. D., Corminboeuf, C., & Wheeler, S. E. (2012). *Journal of Physical Chemistry A, 116*, 3436–3447.
51. Raghavachari, K., Stefanov, B. B., & Curtiss, L. A. (1997). *Journal of Chemical Physics, 106*, 6764–6767.
52. Feller, D., Peterson, K. A., & Dixon, D. A. (2008). *Journal of Chemical Physics, 129*, 204105–204136.
53. Ruscic, B. (2012). Active Thermochemical Tables, early beta 1.110.
54. Lasdon, L. S., & Waren, A. D. (1986). *Generalized Reduced Gradient -2 User's guide (as implemented in the Excell program)*; School of Business Administration, University of Texas at Austin.
55. Rodriquez, C. F., Bohme, D. K., & Hopkinson, A. C. (1996). *Journal of Physical Chemistry, 100*, 2942–2949.
56. Robles, E. S. J., & Chen, P. (1994). *Journal of Physical Chemistry, 98*, 6919–6923.
57. Luo, Y. R., & Benson, S. W. (1997). *Journal of Physical Chemistry A, 101*, 3042–3044.
58. Berry, R. J., Burgess, D. R. F., Nyden, M. R., & Zachariah, M. R. (1995). *Journal of Physical Chemistry A, 99*, 17145–17150.
59. Feller, D., Peterson, K. A., Jong, W. A. D., & Dixon, D. A. (2003). *Journal of Chemical Physics, 118*, 3510–3522.
60. Kormos, B. L., Liebman, J. F., & Cramer, C. J. (2004). *Journal of Physical Organic Chemistry, 17*, 656–664.
61. Zhou, W., Seccombe, D. P., Tuckett, R. P., & Thomas, M. K. (2002). *Chemical Physics, 283*, 419–431.
62. Traeger, J. C., & McLoughlin, R. G. (1981). *Journal of American Chemical Society, 103*, 3647–3652.
63. Dixon, D. A., Grant, D. J., Christe, K. O., & Peterson, K. A. (2008). *Inorganic Chemistry, 47*, 5485–5494.
64. Grant, D. J., Garner, E. B., Matus, M. H., Nguyen, M. T., Peterson, K. A., Francisco, J. S., et al. (2010). *Physical Chemistry A, 114*, 4254–4265.
65. Borkar, S. N., Sztáray, B., & Bodi, A. (2012). *International Journal Mass Spectrometry*.
66. Rodgers, A. S., Chao, J., Wilhoit, R. C., & Zwolinski, B. J. (1974). *Journal of Physical Chemistry Reference Data, 3*, 117–140.

Chapter 5
Photodissociation Dynamics of Four Fluorinated Ethenes: Fast, Slow, Statistical and Non-statistical Reactions

5.1 Preamble

The results presented in this chapter have been previously published as a journal article entitled 'Dissociation dynamics of fluorinated ethene cations: from time bombs on a molecular level to double-regime dissociators' by J. Harvey, A. Bodi, R. P. Tuckett and B. Sztáray, in 2012 in the Royal Society of Chemistry journal, Physical Chemistry Chemical Physics, volume 14, pages 3935–3948. The majority of the data collection and analysis was performed by myself, however, the assistance lent by Ms. Nicola Rogers, Drs. Mathew Simpson Andras Bodi, Melanie Johnson, and Professor Richard Tuckett during beamtime with the collection of the data is gratefully acknowledged. The modelling program was developed by B. Sztáray, A. Bodi and T. Baer [1].

5.2 Introduction

The study of the swift dissociation dynamics ($k > 10^7$ s^{-1}) of small systems, the halogenated methanes dissociating into their first daughter ions, has been presented in Chap. 4. It was found that to accurately locate the E_0, inspection of the breakdown diagram can be an inadequate method, and modelling the breakdown diagram is required. However for such systems, only the internal energy distributions are necessary to model it. In this chapter, attention is also given to the dynamics of slowly dissociating molecules (those on a timescale of 10^3 s$^{-1} < k < 10^7$ s^{-1}). Furthermore, whilst only the appearance energy of the first product daughter ion was measured in those studies within Chap. 4, parallel and sequential reaction channels of the four fluorinated ethenes; $C_2H_3F^+$, 1,1-$C_2H_2F_2^+$, $C_2HF_3^+$ and $C_2F_4^+$ are now explored.

The C–F bond is one of the strongest in organic molecules. Exceptions include the C–H bond in acetylene, the C=C double and C≡C triple bonds [2]. This makes fluorinated alkanes and alkenes particularly appealing subjects in studies of their bonding, electronic spectroscopy and dissociation properties, because the strong bonding also results in sparsely spaced electronic levels. In addition, the small size of the fluorine atom makes these organic compounds amenable to computational chemistry studies, in which thermochemical properties such as enthalpies of formation can be determined [3]. In contrast to saturated perfluorocarbons [4], which photoionize dissociatively even at their ionization energy so that parent ion signal is not detectable, the unsaturated fluorinated ethenes form stable molecular ions [5]. Partly because of this great stability, early studies of fluorinated ethene cations have shown they are metastable with respect to dissociation at low internal energies [6] and can exhibit isolated state behaviour [7, 8].

The dissociative photoionization of monofluoroethene and 1,1-difluoroethene was first investigated using threshold coincidence techniques by Güthe et al. [9], who reported complete kinetic energy release distributions (KERD) for the HF and F loss reaction channels based on the time-of-flight (TOF) spectra of the daughter ions. However, the insufficient mass resolution in the experiment did not allow for the determination of the appearance energy of the F-loss product, $C_2H_2F^+$, from 1,1-$C_2H_2F_2^+$. In a second paper, Güthe et al. [10] further explored the metastable nature of the parent ion in the lowest energy dissociation channel, i.e. HF elimination from both $C_2H_3F^+$ and 1,1-$C_2H_2F_2^+$. Lifetimes on the order of several μs were found using both linear and reflectron time-of-flight mass spectrometers. They reported dissociation rate constants for both ions over a range from threshold to 400 meV above threshold, the smallest of which, $8\bullet10^4$ s^{-1}, was observed with the linear TOF. A tight 4-membered ring transition state with a calculated reverse barrier of 163 kJ mol^{-1} had been suggested for HF loss from 1,1-$C_2H_2F_2^+$ [11], in contrast with the smaller measured reverse barrier of only 95 kJ mol^{-1}. Analogously to HCl loss from C_2H_5Cl [12] or H_2 loss from C_2H_4 [13], Güthe et al. proposed H atom tunnelling to explain this discrepancy. The bimodal behaviour, composed of the statistical as well as the non-statistical dissociations, of F-loss from 1,1-$C_2H_2F_2^+$ was investigated by Gridelet et al., by examination of the KERDs using the maximum entropy method [14]. Only the lower energy dissociative photoionization modus was found to be a statistical adiabatic reaction from the ionic ground state of the parent molecule, which formed a narrow KERD component.

The dissociative photoionization dynamics of trifluoroethene have not previously been studied. Tetrafluoroethene was the subject of a threshold coincidence study by Jarvis et al. [5]. They reported that F loss from $C_2F_4^+$ is accompanied by high kinetic energy (KE) release, too large to be justified by a purely impulsive model, and they suggested two explanations. First, that the heat of formation used to determine the thermochemical threshold for $C_2F_3^+$ production was too high, and dissociation occurs below 15.85 eV. Second, that $C_2F_4^+$ may decay via a 'modified impulsive' mechanism, where energy is deposited exclusively into the rotational and translational modes.

In this chapter, the imaging photoelectron photoion coincidence (iPEPICO) experiment [15] at the VUV beamline [16] of the Swiss Light Source (SLS) was used to prepare and study the dissociation dynamics of internal energy selected ions of monofluoroethene, 1,1-difluoroethene, trifluoroethene and tetrafluoroethene in the 13–20 eV photon energy range with a resolution much higher than in previous studies, i.e. only a few meV. The residence time of photoions in the acceleration region of the TOF mass spectrometer is several µs. If, while the ion resides in the acceleration region, there is significant dissociation then the fragment ion peak shapes are asymmetric and their analysis can yield dissociation rate constants [17], which are measured in the $10^3 \text{ s}^{-1} < k < 10^7 \text{ s}^{-1}$ range. This effect is distinct and different from a symmetrical TOF peak broadening due to kinetic energy release. The iPEPICO experiment yields both the threshold photoelectron spectrum (TPES) as well as parent and daughter ion fractional abundances as a function of the photon energy, which translates into an ion internal energy scan when the ion signal is evaluated in coincidence with threshold electrons. Metastable and parallel fragmentations can be modelled in the framework of the statistical theory of unimolecular reactions: the asymmetric TOF distributions yield the rate curve, $k(E)$, as a function of internal energy which can be extrapolated to the 0 K appearance energy, E_0, where $k(E)$ vanishes. Accurate appearance energies of the daughter ions at 0 K can thus be established even when the low reaction rates result in incomplete dissociation of the parent ions, an effect often referred to as the kinetic shift [18]. For fast dissociations in small molecules, the disappearance energy of the energy-selected parent ion signal yields the 0 K appearance energy, i.e. the energy at which all photoions, including those formed from neutrals with zero internal energy, are above the threshold [19].

What does this appearance energy mean? Most ionic dissociations consist of simple bond breaking, which take place along purely attractive potential energy curves. In such instances, the 0 K appearance energy equals the dissociative photoionization energy, E_{dp}. This E_{dp} value can be used in thermochemical cycles to determine 0 K enthalpies of formation for daughter ions, when the precursor parent enthalpy of formation is known, or vice versa, see Fig. 5.1a [20]. In reactions that involve rearrangements, such as HF-loss, the barrier in the backward direction also has to be considered. Neglecting tunnelling, the appearance energy and the dissociative photoionization energy together can yield the value of this backward or reverse barrier, E_{rb} in Fig. 5.1b. Please refer to Appendix B for all input parameters used to model the breakdown curves.

A process is considered statistical if the complete phase space is accessible to the system. The ion density of states is dominated by the ground electronic state of the parent cation, which implies that the dissociation takes place from this ground electronic state. The adiabatic ionization energies of monofluoroethene, 1,1-difluoroethene, trifluoroethene and tetrafluoroethene are 10.37, 10.30, 10.14 [21] and 10.11 eV [22], respectively. The dissociative photoionization channels all take place above 13 eV at energies corresponding to excited valence states of the four parent cations or in Franck–Condon gaps. If decay processes from these excited states to the ground state are slower than other processes, such as fluorescence or even dissociation, some excited states may have an isolated character and follow a

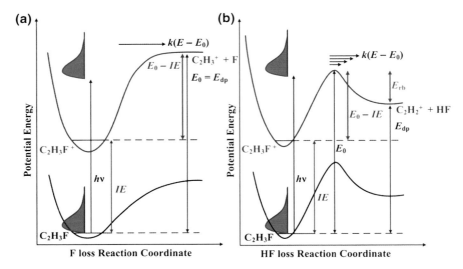

Fig. 5.1 Energy diagram for the dissociations of **a** $C_2H_3F^+$ into $C_2H_3^+ + F$ *without* and **b** $C_2H_3F^+$ into $C_2H_2^+ + HF$ *with* a reverse barrier, IE is the ionization energy, E_{dp} is the dissociative photoionization energy, E_{rb} is the height of the reverse barrier, E_0 is the 0 K appearance energy at which the products are first energetically accessible in the absence of tunnelling, and $E_0 - IE$ is the height of the forward barrier. When there is no reverse barrier present, $E_0 = E_{dp}$

non-statistical path. This has been suggested for several halogen containing ions, such as C_2F_4 [5], CF_3I [23], $SiCl_4$ [24], as well as $Sn(CH_3)_3Cl$, $Sn(CH_3)_3Br$ [25] and even for CH_3OH [26, 27]. There are features uncharacteristic of statistical processes present in the breakdown diagrams of all four fluorinated ethene ions studied in this thesis. Most notably, the fractional abundance of the daughter ions arising from F loss often follows the band intensities of the TPE spectrum of the neutral molecule.

Two intriguing aspects of the dissociative photoionization of fluorinated ethenes are of particular interest to this work. First, we elaborate on the previously observed metastability of the parent ion when HF is lost. The slow dissociation rates correspond to parent ion lifetimes in the μs range, and the large reverse barriers to HF formation lead to impulsive dissociations with more than 1 eV kinetic energy being released. Since the leaving neutral and the fragment ion have comparable masses, a significant portion of this kinetic energy is deposited in the ion and leads to TOF peak broadening. Thus, these metastable parent ions are veritable time bombs with long delays in decay, but with eventual explosive fragmentation. Second, non-statistical dissociations are often associated with impulsive processes occurring on ion surfaces with a strongly repulsive character, as in ground electronic states of CF_4^+ or CCl_4^+ [24, 28], or with fluorescence, i.e. an alternative relaxation pathway, as in N_2O^+ [29, 30]. However, as will be shown in this chapter, this is not always the case; long lived excited electronic states *can*

5.2 Introduction

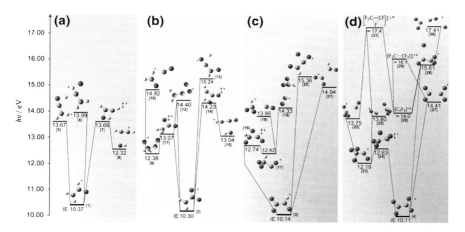

Fig. 5.2 Schematic of the main photoionization dissociation pathways in **a** monofluorethene, **b** 1,1-difluoroethene, **c** trifluoroethene and **d** tetrafluoroethene. Calculated G3B3 values, in eV, are shown for minima and saddle points on the ground electronic state potential energy surfaces. For C_2F_4, the blue plot shows TD-DFT values for the 1st electronic excited state. Continuous lines show observed reactions, dashed lines indicate reactions absent from the dissociative photoionization mechanism. *IE* denotes the experimental adiabatic ionization energy of the parent molecule, also in eV. Transition states are denoted with the superscript ‡

in fact dissociate statistically when only the ground electronic state phase space is inaccessible to the system, and the long lifetimes allow for the statistical redistribution of the internal energy among the nuclear degrees of freedom (Fig. 5.2).

5.3 Results and Discussion

5.3.1 Monofluoroethene

The breakdown diagram for C_2H_3F in the 13–21 eV photon energy range and the modelled breakdown curves of the first two daughter ions together with the experimental points in the photon energy range of 13.2–14.0 eV are shown in Fig. 5.3. The TOF signal for the first daughter ion close to the onset, $C_2H_2^+$, the product of HF loss, has an asymmetric peak shape complete with a long pseudo-exponential tail toward higher times-of-flight, indicating HF loss to be a metastable process (Fig. 5.4). However, even at zero parent fractional abundances, i.e. at energies for which $k(E) > 10^7$ s^{-1}, the $C_2H_2^+$ peak is still broad, but symmetric. This is a consequence of the impulsive nature of HF loss, and the resulting TOF difference between forward and backward scattered ions. The G3B3 calculated reaction enthalpy at 0 K for $CHF=CH_2 \rightarrow HC\equiv CH^+ + HF$ is 12.32 eV (cf. 12.31 eV, based on the heats of formation for $C_2H_3F^+$, −132.2 kJ mol^{-1} [31],

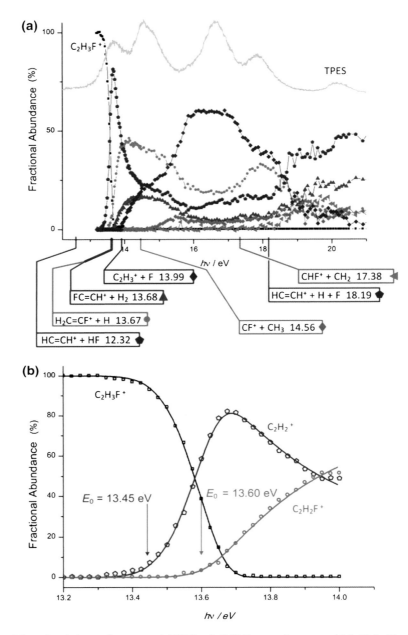

Fig. 5.3 a Breakdown diagram and TPES of C_2H_3F over the range 13.2–21.0 eV. G3B3 calculated onsets at 0 K for selected fragment ions are also included. **b** Modelled breakdown curve (*solid lines*) with experimental points (*open shapes*) for the parent ion $C_2H_3F^+$, and the onsets for only the first two daughter ions, $C_2H_2^+$ and $C_2H_2F^+$, in the energy range 13.2–14.0 eV

$C_2H_2^+$, 1329 kJ mol^{-1} [31, 32], and HF, -273.3 kJ mol^{-1}) [33], whereas for the formation of $H_2C=C^+$ + HF it is 14.12 eV. Therefore the acetylene ion is formed, as reported in an earlier PEPICO study of Dannacher et al. [8]. Our ab initio results also show that the energetically more favourable 1,2-HF elimination proceeds via a tight 4-membered ring transition state involving a H migration across the C=C bond, with a large reverse barrier in the exit channel, see [1] → [7]‡ → [8] in Fig. 5.2a.

By simultaneous fitting of the breakdown diagram and the daughter TOF peaks to obtain the rate curve given in Fig. 5.5a, the experimental 0 K appearance energy for HF loss has been determined to be 13.45 eV. The slow rates seen in Fig. 5.5 are a consequence of the large density of states of the dissociating ion resulting from the large barrier, as well as the small number of states of the tight transition state. Once the system has surmounted this barrier, there is significant excess energy in the reaction coordinate. This energy is not redistributed among the rovibrational modes, causing the fragments $C_2H_2^+$ and HF to fly apart with considerable translational kinetic energy. The experimental 0 K appearance energy and the calculated endothermicity of the dissociative photoionization yield a reverse barrier to HF loss of 1.14 eV. This can be compared with a purely ab initio derived barrier of 1.34 eV, Fig. 5.2a.

The G3B3 calculated onset for H-atom loss, $CHF=CH_2$ → $C_2H_2F^+$ + H + e$^-$, is 13.67 eV when the hydrogen atom is lost from the fluorinated carbon [1] → [5], and 14.71 eV when it is lost from the CH_2 group. The 0 K appearance energy of this daughter ion (m/z 45) is experimentally determined to be 13.60 eV, suggesting that the former hydrogen atom loss process giving rise to $C_2H_2F^+$ is not kinetically hindered. Indeed, no reverse barrier to hydrogen atom loss could be found in our calculations, thus the metastable decay close to threshold, see rate data in Fig. 5.5a, is mostly due to the large barrier and the correspondingly large density of states in the parent ion. The observation of this metastability supports results reported by Güthe et al. [9]. Since the H-loss transition state is looser than the HF-loss one, the competition between the first two channels favours the former, with the $C_2H_2F^+$ fractional abundance some 20 % higher than that $C_2H_2^+$ between 14 and 18 eV. Above 18.4 eV the loss of 20 a.m.u. becomes the dominant channel. This is identified as the formation of $HC=CH^+$ + H + F + e$^-$, for which the G3B3 calculated onset is 18.19 eV. The reaction endothermicity of $CHF=CH_2$ → $H_2C=C^+$ + H + F + e$^-$ is calculated to be 19.99 eV, and is subsequently discounted as the origin of the signal below 20 eV. Consequently the m/z 26 daughter ion $C_2H_2^+$ is derived from the sequential dissociation of $C_2H_3^+$ by H loss as well as from $C_2H_2F^+$ by F loss in this energy range.

From their threshold to about 0.5 eV above, the fractional ion abundances of C_2HF^+ and $C_2H_3^+$ rise less steeply than those of the first two daughter ions, $C_2H_2^+$ and $C_2H_2F^+$. The appearance energy, AE, of C_2HF^+ and $C_2H_3^+$ are measured to be 13.7 and 13.9 eV, respectively. The thermochemical onset for 2,2-H_2 elimination yielding $FHC=C^+$ is calculated to be 15.62 eV and cannot take place in this energy range. Therefore the structure of C_2HF^+ must be $CF=CH^+$, which is confirmed by the calculated 1,2-H_2 elimination threshold of 13.68 eV. Contrary to

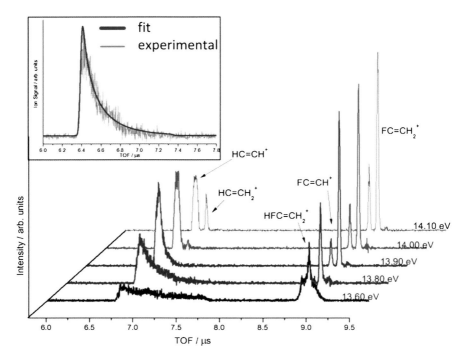

Fig. 5.4 Selected time-of-flight distributions for C_2H_3F in the 13.6–14.1 eV photon energy range. The parent ion is observed at 8.8 μs and the first HF-loss daughter fragment HC=CH$^+$ at 6.6 μs. The asymmetric peak shape is a consequence of slow dissociation in the acceleration region. The $C_2H_2F^+$ ion due to metastable H loss, is also seen in the 8.7–8.8 μs range as a shoulder to the parent peak. At higher energies the formation of C_2HF^+ and $C_2H_3^+$, due to H_2 and F loss, are clearly seen in the 8.6–8.7 and 6.7–6.8 μs TOF ranges, respectively. Above 14 eV, the kinetic energy release in the HC=CH$^+$ ion is evident in a broadened peak. Inset is the modelled TOF fit (*thicker line*) for the metastable peak of HC=CH$^+$, associated with HF loss at 13.70 eV

1,1-difluoroethene, in which only 2,2-H_2 elimination is structurally possible, H_2 loss can compete effectively with the other dissociation channels in monofluoroethene. The agreement between the calculated and the experimental onsets also suggest that H_2 loss is not slow at threshold, quite unlike HF loss. This is only possible if H_2 loss has no reverse barrier along the reaction coordinate, or if it is very narrow and there is fast tunnelling through it.

The mechanism of F loss yielding $C_2H_3^+$ has been discussed extensively in the literature [8, 9, 34, 35]. This process is observed at its thermochemical threshold, and its rise is consistent with a statistical competitive fast reaction with a loose transition state. As can be seen in the TOF distributions (Fig. 5.5), the parent ion ceases to be metastable in this energy range and the F-loss signal is readily identified in our experiment, in contrast to a previous report [9]. However, at 15.5 eV, there is a sudden increase in the $C_2H_3^+$ abundance which fits poorly into the statistical picture. Previously, it was proposed that isolated $\widetilde{C}$ state behaviour

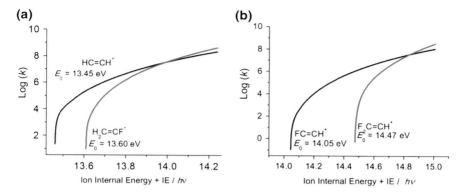

Fig. 5.5 Plot of $\log_{10} k$ (E) versus $h\nu$ for **a** HF loss and H loss from $C_2H_3F^+$, **b** HF loss and F loss from 1,1-$C_2H_2F_2^+$. The experimental rates, observed in the 10^3 s^{-1} < k < 10^7 s^{-1} range, are extrapolated to obtain the E_0

(i.e. the dissociation dynamics are dominated by those of the electronic $\widetilde{C}$ state of the parent ion) contributes to this signal [8, 34, 35]. However, the $\widetilde{C}$ peak in the TPES is observed at an onset of 16.18 eV, whereas this sudden rise occurs some 0.7 eV lower, still in the energy range of the $\widetilde{B}$ peak. Furthermore, the $C_2H_3^+$ ion abundance follows the $\widetilde{C}$ peak only very approximately. Consequently, we confirm the double nature of the F-loss process, but also suggest that the $\widetilde{C}$ state is not playing a simple and direct role in the non-statistical range. Instead of $\widetilde{C}$ state participation, it is more likely that Rydberg series converging to the $\widetilde{C}$ state have different autoionization pathways leading to the $C_2H_3^+$ product. Based on the KER analysis of the $C_2H_3^+$ ion yield at 16.85 eV, Momigny and Locht [35] conclude that approximately two-thirds of the ion flux dissociates on the $\widetilde{C}$ state producing the electronically excited $\widetilde{a}\ ^3A''$ state of $C_2H_3^+$, which can then internally convert to its ground state, thereby keeping most of the excess energy. However one-third of the ion flux arrives at the ground state of $C_2H_3F^+$, which correlates with the ground $\widetilde{X}\ ^1A'$ state of $C_2H_3^+$, allowing for a larger kinetic energy release. Such a bimodal behaviour has also been proposed by Gridelet et al. for the F-loss pathway from 1,1-$C_2H_2F_2^+$ [14]. Indeed, there is a very swift decrease in half the $C_2H_3^+$ signal together with a jump in the $C_2H_2^+$ fractional abundance at around $h\nu = 18.5$ eV. Taking into account the 0 K heats of formation of $C_2H_3^+$ [36], $C_2H_2^+$ [31, 32], and H [37], (1,120, 1,329 and 216 kJ mol^{-1}, respectively), $C_2H_3^+$ is expected to lose a further H atom at an internal energy of 4.4 eV, i.e. at a photon energy of 18.4 eV, whilst the G3B3 value for the dissociative photoionization $C_2H_3F \rightarrow HC=CH^+ + F + H + e^-$ is 18.19 eV. As will be shown later for C_2F_4, the breakdown diagram of a sequential dissociation corresponds to the internal energy distribution in the first dissociation step, and can be used to study the excess energy redistribution. Thus, we attempted to analyse the $C_2H_3^+$ versus

$C_2H_2^+$ breakdown curves in the 18–19 eV range to determine the $C_2H_3^+$ internal energy distribution. There is a difference of about 1 eV in the excess energy available for kinetic energy release depending on whether the excited or ground state $C_2H_3^+$ intermediate is formed. Both pathways yielded an acceptable fit to the $C_2H_2^+$ breakdown curve within the signal-to-noise ratio of the experimental data. Ergo, the comparatively noisy high-energy breakdown curves of the three different open channels (H + F loss, F + H loss with an $\widetilde{X}\,^1A'$ or $\widetilde{a}\,^3A''$ $C_2H_3^+$ intermediate) and the small differences in their energies (1.3 and 1 eV more excess energy available for KER in the first two) do not allow for a sufficiently detailed description of the reaction mechanism yielding $C_2H_2^+$.

CF^+ (m/z 31) appears around a photon energy of 14.87 eV, which is 0.3 eV higher than the G3B3 calculated endothermicity for $CHF=CH_2 \rightarrow CF^+ + CH_3 + e^-$, 14.56 eV. It is 0.17 eV higher than the previously reported thermochemical value of 14.704 eV [9] and lies between previous appearance energies of 14.5 [8] and 14.90 eV [9]. Methyl radical loss is preceded by H atom migration, and ab initio calculations were used to obtain a plausible pathway to CF^+ production. The transition state to $CF-CH_3^+$ was calculated to lie at 12.07 eV, well below the overall barrier to CF^+ formation. The highest energy major channel observed in this work is C=C bond cleavage to form $CHF^+ + CH_2$. It has a calculated onset energy of 17.38 eV and is seen experimentally at 18.4 eV. This value is ca. 2 eV lower than the appearance energy of 20.02 eV reported by Güthe [9]. The thermochemical threshold to $CHF^+ + CH_2$, 17.099 eV [38], is in reasonable agreement with our calculated G3B3 value, confirming the competitive shift in the CHF^+ signal. At such high internal energies numerous processes can take place at rates comparable to intramolecular vibrational relaxation. Therefore, the fact that we observe a further parallel channel opening up at all is remarkable.

5.3.2 1,1-Difluoroethene

The breakdown diagram of 1,1-$C_2H_2F_2^+$ in the 13.9–21.0 eV energy range with ab initio dissociative photoionization energies for selected channels, as well as the experimental and modelled breakdown curves for the HF and F loss reactions in the 13.9–14.7 eV energy range, are shown in Fig. 5.6. Similarly to monofluoroethene, HF loss is the lowest energy channel and the G3B3 calculated endothermicity lies 1.4 eV lower than that for F-atom loss. The calculated reaction energy for $F_2C=CH_2 \rightarrow FC=CH^+ + HF + e^-$ at 0 K is 13.04 eV and the experimental 0 K onset energy, obtained by simultaneous modelling of the breakdown diagram and the daughter ion TOF spectra, is 14.05 eV. This agrees with the value of 14.1 eV reported by Güthe [9], and indicates a reverse barrier of 1.01 eV in [**2**] → [**14**]‡ → [**15**]. The daughter ion TOF peak shapes indicate metastable behaviour, and our calculations predict a tight transition state. The purely calculated E_{rb} of 1.19 eV is, as for $C_2H_3F^+$, somewhat higher than the value based on the experimental E_0. This small discrepancy of 0.18 eV could be explained by tunnelling through the reverse barrier, which effectively lowers the observed E_0. Our values agree with the previously measured E_{rb} of 0.98 eV but not with the reported ab initio value of 1.69 eV [10, 11]. This indicates that most of the reported 0.71 eV difference was primarily due to the inadequate description of the potential energy surface at the UHF/6-31G(d)//UHF/STO-3G level of theory.

The calculated onset energy for the formation of $C_2H_2F^+$ (m/z 45) by F loss is 14.40 eV. The corresponding breakdown curve, however, is noisy due to the background subtraction required because the large asymmetric TOF signal of $FC=CH^+$ overlaps with the $FC=CH_2^+$ signal from F-loss (Fig. 5.7). We performed a potential energy scan along the C–F bond stretch coordinate to obtain the potential energy curve for F-atom loss. Figure 5.2b shows that a transition state at a C–F bond length of 1.8 Å is predicted $[11]^{\ddagger}$, in which the leaving fluorine atom straddles the C=C bond. This transition state may lead either to F-loss (in which there is no overall reverse barrier) or to the $CH_2F–CF^+$ isomer ion [9].

F-loss may proceed without encountering this transition state, and this path is selected for the modelling of the dissociation rates. Figure 5.6b shows the breakdown curve modelling, which led to the F-loss 0 K appearance energy of 14.47 ± 0.1 eV. As previously observed by Güthe et al. [9], $C_2H_2F^+$ [12] is the most abundant daughter ion between 16 and 17 eV as a result of a non-statistical process. As with C_2H_3F, there appears to be two pathways at play. At lower energies, F loss is a statistical process on the ionic ground state potential energy surface, but quickly loses out to reactions involving CF and CH_2F loss above 14.7 eV. The diminishing $C_2H_2F^+$ fractional abundance starts rising again around 15.3 eV, in coincidence with the onset of the $\widetilde{B}$ state in the TPES. This apparent similarity is indicative of isolated-state, non-statistical decay from this state of $C_2H_2F_2^+$. However, as is the case for monofluorethene, the breakdown curves only approximately follow the TPES, indicating a complex mechanism.

The G3B3 onset energies for CH_2F^+ and CF^+ are close to one another at 14.82 and 14.92 eV, and their experimental onsets are 14.70 and 14.86 eV, respectively. These daughter ions are the products of the same process with the charge localised on one or the other fragment. The CF^+ and CH_2F^+ fragments are formed in competition with fast F and HF loss, suggesting a loose transition state and no overall reverse barrier to dissociation. As already mentioned, in the transition state structure the F atom can move over and attach to the CH_2 group in $[11]^{\ddagger}$, $[2] \rightarrow [11]^{\ddagger} \rightarrow [9] \rightarrow [10]$. C–C bond rupture in $[11]^{\ddagger}$ can also lead to CF^+. The ionization energy (*IE*) of CF has been determined by Dyke et al. to be 9.11 ± 0.01 eV [39], whereas that of CH_2F is reported to be 9.04 ± 0.01 eV by Andrews et al. [40]. In the absence of a competitive shift, the offset in onset values would correspond to the ionization energy difference. If there is a competitive shift, i.e. the CF^+ signal is delayed and rises only at higher energies because it is outcompeted by the other parallel channels, this offset can only be considered as an upper limit to the ionization energy difference. Both quantities appear to be well established; hence in lieu of a detailed kinetic model, only a lower limit to the *IE*

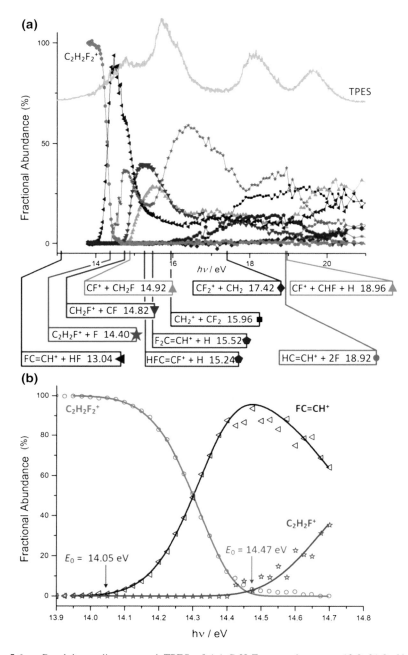

Fig. 5.6 **a** Breakdown diagram and TPES of 1,1-$C_2H_2F_2$ over the range 13.9–21.0 eV. The G3B3 calculated onsets at 0 K for selected fragment ions are also included. **b** Modelled fit (*solid line*) with experimental points (*open shapes*) for the parent ion, 1,1-$C_2H_2F_2^+$, and the onsets for the first two daughter ions, FC=CH$^+$ and $C_2H_2F^+$ in the 13.9–14.7 eV energy range

5.3 Results and Discussion

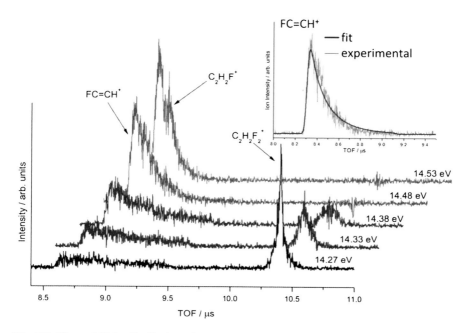

Fig. 5.7 Time-of-flight distributions for 1,1-$C_2H_2F_2$ from the parent ion, at 10.4 μs to the fragment, FC=CH$^+$, at 8.8 μs. The asymmetric peak shape of the daughter ion is a consequence of slow dissociation in the acceleration region. The fast F-loss daughter peak, $C_2H_2F^+$, is seen emerging from the metastable FC=CH$^+$ peak as the energy increases and is found at 8.7 μs. Inset shows the TOF fit for the metastable FC=CH$^+$ peak, at 14.49 eV

of CH_2F is given as 8.95 eV. At energies above 15.3 eV, the signal for these two ions decreases because the non-statistical F-loss channel is preferred.

The calculated onsets of the H-loss products (m/z 63), HFC=CF$^+$ and F_2C=CH$^+$, are 15.24 and 15.52 eV, respectively. Experimentally, H-loss product appears only at a higher photon energy of ca.15.9 eV, primarily because it is outcompeted by the other fast processes at lower energies. This also leads to its slow rise with increasing $h\nu$. At an energy $\sim$1 eV lower than in monofluoroethene, cleavage of the C=C bond occurs. The calculated onset for production of $CH_2^+ + CF_2$ is 15.96 eV, and its experimental appearance energy is 16.9 eV. By contrast, the CH_2 loss is calculated at 17.42 eV but is not seen experimentally until 18.9 eV. The faster rise of CH_2^+ than that of $C_2HF_2^+$ from H loss suggests a looser transition state for the C=C bond rupture. Unlike monofluoroethene, however, the positively charged fragment first seen resulting from C=C cleavage is not the fluorine-containing moiety, but CH_2^+. This observation is explained by the 1.3 eV difference between the *IE* of these fragments (CH_2 [41] 10.39 ± 0.01 eV, CHF [42] 10.06 ± 0.05 eV, CF_2 [43] 11.36 ± 0.005 eV).

Based on energetics considerations, the second rise in the CF$^+$ signal at 19 eV is suggested to stem mostly from the C–C bond cleavage in the H-loss fragment ion,

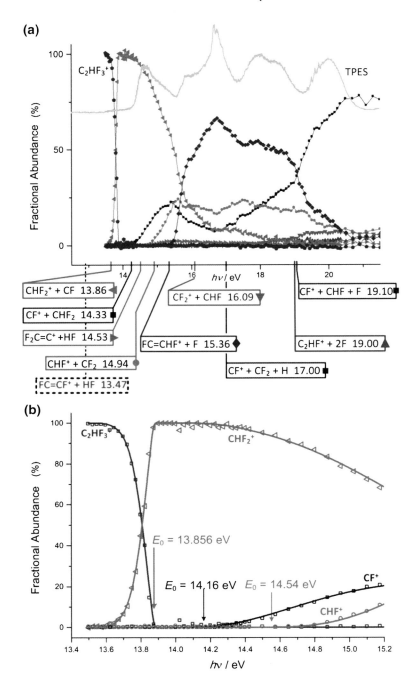

◄ **Fig. 5.8 a** Breakdown diagram and TPES of C_2HF_3 over the range 13.45–21.5 eV. The G3B3 calculated onsets at 0 K for selected fragment ions are also included. The calculated onset for the 1, 2-HF abstraction at 13.47 eV denoted by the black dashed line, is included for reference, though the product ion is not seen experimentally. **b** Experimental points (*open shapes*) with modelled breakdown curve (*solid line*) for $C_2HF_3^+$, and the onsets for the first three daughter ions, CHF_2^+, CF^+ and CHF^+ in the energy range 13.50–15.25 eV

HFC=CF$^+$ (calculated onset is 18.96 eV). This is supported by a decrease in the $C_2HF_2^+$ abundance in this energy range, i.e. a decrease in the H-loss signal. Finally, the decrease in the F-loss signal $C_2H_2F^+$ between 19 and 20 eV is due to two possible consecutive reactions from $C_2H_2F^+$: a further H-loss to FC=CH$^+$ (18.91 eV), or, after a rearrangement to HFC=CH$^+$, a loss of F to HC=CH$^+$ (18.92 eV) in agreement with the mechanism suggested by Güthe et al. [9].

5.3.3 Trifluoroethene

To the best of our knowledge, this is the first report of the fragmentation pathways of ionized trifluoroethene by coincidence techniques. The breakdown diagram and the threshold photoelectron spectrum in the 13.5–21.5 eV photon energy range are shown in Fig. 5.8a. The adiabatic ionization energy of the neutral molecule has been determined to be 10.14 eV [21]. The lowest-energy G3B3 calculated 0 K dissociative photoionization energy of 13.47 eV corresponds to the 1,2-HF elimination. In contrast to monofluoroethene and 1,1-difluoroethene, this reaction is not observed and $C_2F_2^+$ is virtually absent in the breakdown diagram.

The first observed daughter ion, CHF_2^+ corresponds to the loss of the CF fragment which requires an initial F-atom migration [3] → [16]‡ → [18]. The G3B3 calculated onset energy is 13.86 and eV the experimental 0 K appearance energy is measured to be 13.856 ± 0.007 eV, so there appears to be no reverse barrier in the exit channel. Figure 5.8b shows the experimental data, the modelled breakdown diagram, and the obtained 0 K appearance energies determined for the first three dissociation channels. The TOF peaks corresponding to CHF_2^+ are almost symmetric, so we conclude that the parent ion is barely metastable along this reaction coordinate and dissociation is therefore fast. When a rearrangement precedes the loss of HF in ionized mono- and 1,1-difluoroethene, these reactions have slow rate constants. Therefore it might seem counterintuitive that the rates for CF loss from ionized trifluoroethene are not slow. To shed light on this issue, we obtained ab initio potential energy curves leading to these fragments, see Fig. 5.2c. The F-transfer transition state in this series has a 3-membered ring structure [11]‡, [16]‡ and [24]‡ whereas HF-loss proceeds via a 4-membered ring transition state structure [7]‡ and [14]‡. The figure also shows that the reaction coordinate changes character as the reaction proceeds. Initially, it starts as a F-atom migration across the C=C bond leading to a HF_2C-CF^+ minimum, but this is followed by a C–C bond cleavage to form CHF_2^+ + CF, [3] → [16]‡ → [17] → [18]. The F-transfer transition state [16]‡ lies at 12.74 eV and the H-transfer transition state has been found to lie at 12.93 eV, so both pathways are possible, though the lower energy F-transfer is more favourable. The reverse barrier associated with the F-migration is much smaller than the dissociation endothermicity, so there is no overall reverse barrier to production of $CHF_2^+ + CF$ or $CF^+ + CHF_2$. Thus, the 0 K appearance energy of CHF_2^+, 13.856 ± 0.007 eV, corresponds to the dissociative photoionization energy.

The second daughter ion observed is CF^+, corresponding to the loss of the CHF_2 fragment, [3] → [16]‡ → [17] → [19]. The experimentally determined 0 K appearance energy for this ion is 14.16 ± 0.02 eV, with the G3B3 onset energy calculated to be 14.33 eV. As these first two dissociative photoionization reactions differ only in which moiety the positive charge is localized on, the difference in the E_0 values, 0.30 ± 0.02 eV, yields the difference in the ionization energies of the CF and CHF_2 radicals. The ionization energy of CF is well established, 9.11 eV ± 0.01 [39], whilst values for CHF_2 span a large range of experimental values, 8.78 [38], 8.74 [44], and 10.5 eV [45], and a calculated value of 8.4 eV [46]. By anchoring to the CF value, we determine the *IE* of the CHF_2 radical to be 8.81 ± 0.02 eV. The abundance of CF^+ has two maxima, the first at ca. 15.3 eV (fractional abundance of 25 %) and a much larger one at ca. 20.5 eV (80 %). The shape of its breakdown curve can help understand its production mechanism. At low energies, CF^+ is produced by the $HFC=CF_2 + h\nu \rightarrow CF^+ + CHF_2 + e^-$ reaction. At 17 eV, a new channel opens up in which the third daughter ion, CHF^+, which is produced initially by C=C bond cleavage to form $CHF^+ + CF_2$, loses an H atom in a sequential process to produce CF^+. At 19.1 eV, the steepness of the CF^+ yield further increases as C=C bond rupture becomes possible from the F-loss daughter ion $CHF=CF^+$.

The 0 K appearance energy of the third daughter ion, CHF^+ is 14.54 ± 0.02 eV. G3B3 calculations give 14.94 eV, corresponding to cleavage of the C=C bond of the parent ion, with CF_2 as the neutral fragment, [3] → [21]. This process occurs at relatively low photon energies for trifluoroethene and is an example of the perfluoro-effect [47], i.e. a decrease in the C=C bond strength as the number of F substituents increases. The onset for CH_2^+ production from 1,1-difluoroethene is at ca. 17 eV, the onset of CHF^+ from monofluoroethene does not occur until 18 eV. For all three molecules, however, this never becomes a dominant channel, with the maximum fractional abundance (CH_2^+ from 1,1-difluoroethene) never exceeding 35 %.

The final major fragment ion formed from trifluoroethene is F-loss to $HFC=CF^+$, [3] → [20]. Its appearance energy is 15.36 eV and it turns on at its calculated thermochemical threshold. This reaction channel is associated with non-statistical F-loss, because the F-loss curve increases too sharply over a narrow energy range to be statistical. This channel is the most abundant yield between 15.6 and 19.0 eV, and the signal emulates closely that of the TPES. This range of energies coincides with the $\widetilde{E}\big/\widetilde{F}\big/\widetilde{G}$ excited states of the ion where ionization occurs from C–F orbitals [21]. Unlike the F-loss channel observed from $C_2H_3F^+$

and 1,1-$C_2H_2F_2^+$, the contribution of the statistical F loss is not seen and is suggested to be minor at all energies. Higher-energy channels with abundances less than 25 % occur after these four major channels: these are the production of CF_2^+ + CHF at 17.2 eV and C_2HF^+ + 2 F at 18.9 eV. G3B3 dissociative photoionization energies for these channels have been calculated to be 16.09 and 19.00 eV, respectively.

5.3.4 Tetrafluoroethene

The first three dissociative photoionization channels of C_2F_4 open in a Franck–Condon gap, as shown in the breakdown diagram and threshold photoelectron spectrum, Fig. 5.9a. This observation is in agreement with the findings of an earlier TPEPICO study by Jarvis et al. [5]. The first channel, formation of CF_3^+ with CF as the accompanying neutral, has a calculated onset energy of 13.75 eV. Surprisingly, although analogously to the C_2HF_3 system, the CF_3^+ TOF peak is symmetric and narrow, therefore the fluorine migration and subsequent C–C bond cleavage is a fast process without a large reverse barrier. At somewhat higher energies, CF^+ is the second daughter ion, again mirroring the second dissociative photoionization channel in trifluoroethene. The adiabatic ionization energy of C_2F_4 is 10.11 ± 0.01 eV [22], meaning that the total depth of the potential energy well to CF_3^+ + CF is about 3.64 eV, leading to a high density of states in the dissociating ion. In contrast with trifluoroethene, no reasonably chosen transition state is loose enough to lead to rates larger than 10^7 s^{-1} at such high internal energies.

This indicates that the F-transfer mechanism plays a crucial role in ensuring that there is no kinetic shift. Rearrangement to a CF_3CF^+ intermediate, [4] → [24]‡ → [23], can take place at a much lower energy than the E_0 of CF_3^+. Even though the transition state for this process is relatively tight, the rates are fast at an excess energy of 1–2 eV, i.e. at the dissociative photoionization onset: ab initio RRKM rates, based on the G3B3 calculated transition state are in excess of 10^9 s^{-1} at threshold. C–C bond rupture can then proceed through a loose transition state with a lower density of states in the dissociating intermediate, giving rise to fast rates and no kinetic shift for [4] → [24]‡ → [23] → [22] or [25]. In the absence of this CF_3CF^+ intermediate [23], the dissociation would be slow and a kinetic shift apparent in the spectrum. Figure 5.9b shows the modelled breakdown curves, and the E_0 for CF_3^+ production has been determined to be 13.717 ± 0.007 eV.

The appearance energy of CF_3^+ and that of the second daughter, CF^+, are very close, as the ionization energy of CF_3, somewhat controversially reported as 8.61 eV [48], 8.6–8.7 eV [49], 9.04 eV [50], 9.05 ± 0.004 eV [51], 9.02 ± 0.03 eV and 9.08 ± 0.03 eV [52], is only slightly lower than that of CF, 9.11 ± 0.01 eV [39]. The E_0 to CF^+ + CF_3 formation is determined to be 13.740 ± 0.010 eV based on the statistical modelling. The difference between the barriers to these two daughter ions is 0.023 eV, and, together with the *IE* of CF, the

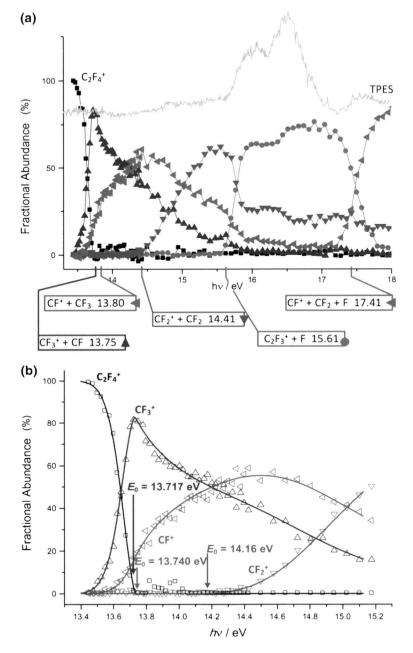

Fig. 5.9 a Breakdown diagram and TPES of C_2F_4 recorded over the range 13.5–18.0 eV. The G3B3 calculated onsets at 0 K for selected fragment ions are also included. **b** Experimental points (*open shapes*) with modelled breakdown curve (*sold line*) for the parent ion, $C_2F_4^+$, and the onsets for the first three daughter ions, CF_3^+, CF^+ and CF_2^+ in the energy range 13.4–15.2 eV

IE of CF_3 is determined to be 9.090 ± 0.015 eV. This lies towards the higher end of previous reported onset values. While our CF_3^+ E_0 agrees well with the result from the photoionization efficiency (PIE) curve of C_2F_4 of 13.721 ± 0.005 eV [51], our CF^+ onset differs considerably from the value of 13.777 ± 0.005 eV reported by Asher and Ruscic [51]. Presumably, the reason is that the competitive shift in the CF^+ channel was not considered in the PIE work, leading to a higher reported value. As a consequence, an onset energy difference (0.055 ± 0.003 eV) was reported, which corresponds well to the offset in breakdown curves we observe, but not to the E_0 difference. Thus, we feel that the C_2F_4 photoionization experiment warrants a revision of the IE of CF_3 to 9.090 ± 0.015 eV.

The third channel is formation of CF_2^+ + CF_2, with a calculated appearance energy of 14.41 eV. This reaction arises from cleavage of the C=C bond, [4] → [27], and occurs at a lower photon energy than the same process in trifluoroethene, due to the perfluoro effect [47]. This parallel channel is in competition with the first two channels. The rate curves were obtained for CF_3^+ and CF^+ formation based on the density of states of intermediate [23], which were then used in conjunction with a rate equation based on the parent ion [4] density of states to describe C=C bond breaking. This approach yields a 0 K appearance energy of CF_2^+ of 14.16 ± 0.04 eV.

There is a sharp increase in the abundance of the fourth channel, F atom loss and production of $C_2F_3^+$, at 15.56 eV, at the end of a substantial Franck–Condon gap. Unlike the first three channels, a straightforward statistical treatment is not appropriate for this non-statistical process, because, similarly to F loss from $C_2HF_3^+$, the breakdown curve rises too steeply to be statistical [53]. Two pieces of evidence stand out. First, there is an excellent correlation between the peaks in the TPES and the breakdown curve here. Second, the steepness of the crossover region is not only inconsistent with a parallel competing channel, but, as will be shown later, also corresponds to the room temperature internal energy distribution of $C_2F_4^+$; significantly unlike a crossover due to a slowly changing rate constant ratio of competing statistical processes. The overall breakdown diagram appears to be comprised of two separate outcomes or regimes. The first one consists of the CF_3^+ + CF, CF^+ + CF_3 and CF_2^+ + CF_2 channels discussed so far, which arise from dissociations on the ground state surface of the parent ion, $C_2F_4^+$, partly through the intermediate structure CF_3CF^+. Below a photon energy of 15.5 eV, only reactions belonging to this first regime are observed. Above this energy, a regime change occurs, and the two of the observed reactions belong to the excited, isolated-state second regime: loss of a fluorine atom yielding $CFCF_2^+$ + F, which is followed by the sequential formation of CF^+ + CF_2 + F above 17.2 eV (Fig. 5.10).

As seen in the breakdown diagram in the range 15.9–18.0 eV, regime-two reactions dominate the regime-one reactions by a constant factor of roughly 2:1. The threshold photoionization mechanism is suggested to play a vital role and can be discussed in the framework proposed for iodomethane previously [54]. Following photoabsorption, the neutral C_2F_4 molecule is excited to a Rydberg state with favourable Franck–Condon factors. Three non-radiative decay pathways are

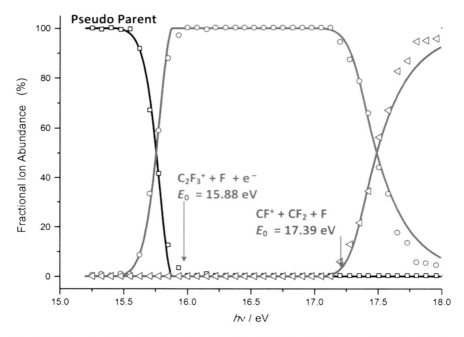

Fig. 5.10 Experimental points (*open shapes*) with modelled breakdown curve (*solid line*) for the regime 2 of the breakdown diagram, of $C_2F_4^+$. As ions formed through regime 2 ($C_2F_3^+$ and CF^+) are decoupled from those of regime 1, all previous ion abundances are grouped together to form the pseudo-parent-ion abundance

possible: (1) crossing to a repulsive neutral curve leading to neutral fragments, which are not detected in our experiment, (2) whilst on this repulsive surface, the system can return to the ground state Rydberg manifold eventually leading to the ground state parent ion which dissociates via regime one, and (3) direct autoionization to an excited electronic state, in this case the $\tilde{A}$ state of $C_2F_4^+$, which dissociates according to regime two by F loss and then by consecutive CF_2 loss.

Here we discuss three aspects of the double-regime dissociation mechanism of $C_2F_4^+$. First, in Fig. 5.2d, TD-DFT potential energy levels are shown for production of $CF_3^+ + CF$, the simple C=C bond breaking and the F-loss channels. Excited state potential energy curves were obtained along the minimum energy path for the ground electronic state, and the TD-DFT minima and maxima are reported here. These points are therefore not necessarily stationary points on the excited state surface, but we believe they are reasonably good representations of them. Some EOM-UCCSD/aug-cc-pVTZ calculations along these curves showed the same general characteristics with only small differences in excitation energy. Since the C–F bond is very strong, F loss cannot compete effectively on the ground electronic surface; the dynamics are dominated by the $CF_3^+ + CF$, $CF^+ + CF_3$ and $CF_2^+ + CF_2$ exit channels. However, if the first excited state is only weakly coupled to the ground state, which is hardly surprising given the 4–6 eV gap

5.3 Results and Discussion

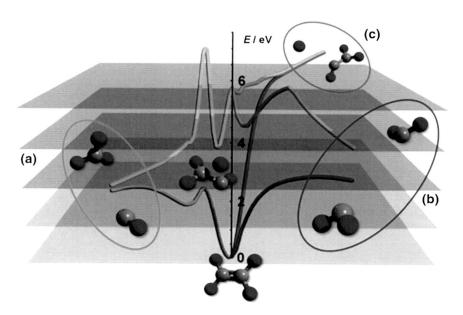

Fig. 5.11 TD-DFT calculations of the three dissociation pathways from the ground electronic state of $C_2F_4^+$, at $E = 0$ **a** $C_2F_4^+ \rightarrow CF_3^+ + CF_2$, $C_2F_4^+ \rightarrow CF_3 + CF_2^+$, **b** $C_2F_4^+ \rightarrow CF_2^+ + CF_2$ and **c** $C_2F_4^+ \rightarrow C_2F_3^+ + F$ at the EOM-UCCSD/aug-cc-pVTZ level. At $E = 5.5$ eV the $\tilde{A}$ state becomes energetically accessible, but only the (**c**) pathway is not blocked by large barriers

between the two states, F loss, [26]* → [28], becomes possible. This is not because the $\tilde{A}$ state converges to energetically disallowed excited state products, as was invoked in the non-statistical model for $Sn(CH_3)_3X$ [25] and methanol [27], but because of large reverse barriers for the other competing processes on the excited state surface. Thus, the three regime-one exit channels, [26]* → [31]‡ ↛ [22], [26]* → [31]‡ ↛ [25] and [26]* → [29]‡ ↛ [27], are kinetically 'blocked' on the $\tilde{A}$ state surface, see Fig. 5.11.

Second, the narrow width of the regime crossover at 15.9 eV corresponds to the width of the thermal energy distribution of $C_2F_4^+$. This observation prompted us to consider the regime-two processes independently of the preceding channels, and plot a 'regime-two breakdown diagram' as shown in Fig. 5.10. This was achieved by disregarding regime-one product ions at $h\nu > 16$ eV, then re-normalizing the signal so that the F-loss daughter ion, $C_2F_3^+$, converges to 100 % closely above 16 eV.

This is done to obtain a regime-two 'pseudo parent ion' signal. The temperature of the internal energy distribution in the 'pseudo parent' that gives the best fit is 340 K, somewhat higher than room temperature. Consequently, in contrast with CH_3I where autoionization to the electronically excited state was found to be enhanced at low internal energies [54], here we find that the direct autoionization process (3) appears slightly enhanced at high internal energies and the thermal energy distribution is somewhat widened. Note that regime-two ions are only distinct from regime-one ions above the F-loss threshold of 15.5 eV. Below this energy, the long-lived electronically excited parent ions will eventually undergo internal conversion to the ground ion state and dissociate to fragment ions via regime one.

Third, in a particularly serendipitous turn, $C_2F_3^+$, produced by non-statistical F-loss, undergoes a further sequential dissociation above 17 eV to form CF^+, [**26**]* → [**28**] → [**30**]. However, sequential F-loss from the regime-one CF_2^+ could interfere with the regime-two CF^+ signal, as evidenced by the small rise in the CF^+ abundance close to 18 eV. Due to the larger kinetic energy release and the different product energy partitioning meaning that more than half of the excess energy is lost in the first C=C bond rupture step, this regime-one process will be very slow to rise with increasing photon energy, confirmed by the almost constant CF_2^+ abundance above 17.5 eV. Therefore, regime-two processes are virtually distinct from regime-one processes.

Repulsive surfaces and impulsive mechanisms are often invoked to explain effective competition between non-statistical and statistical channels [55]. It was only recently that some evidence has been published highlighting the statistical redistribution of internal energy which is possible in isolated-state processes [27]. The breakdown curve of a sequential dissociation yields the product energy distribution of the dissociating ion [17, 25]. By modelling the second step in the regime-two breakdown diagram in Fig. 5.10, it becomes evident that it is *only* the electronic ground state phase space which is inaccessible to our system, and statistical redistribution of the excess energy among the nuclear degrees of freedom can indeed occur. The derived E_0 values of 15.88 ± 0.03 eV and 17.39 ± 0.06 eV for $C_2F_3^+$ + F and CF^+ + CF_2 + F, respectively, can be compared with the G3B3 calculated dissociative photoionization energies of 15.61 and 17.41 eV. The enthalpies of formation for C_2F_4, CF^+, CF_2 and F as listed in Table 5.1 yield an onset for C_2F_4 → CF^+ + CF_2 + F of 17.32 ± 0.06 eV supporting the E_0 derived in this work. This agreement is excellent, confirming the validity of the 'pseudo-parent assumption' and the applicability of the statistical approach to regime two. To summarize, the internal energy distribution of the F-loss daughter ion, $C_2F_3^+$, determines its breakdown curve in the sequential CF_2-loss process. The latter is very well described assuming a statistical redistribution of the excess energy in the F-loss step. Therefore the F-loss is found to be non-statistical only in the sense that the ground electronic state is inaccessible. The statistical approximation is valid for the nuclear degrees of freedom, and F loss is not an impulsive process as was previously proposed [5].

5.3 Results and Discussion

Table 5.1 Thermochemical values (in kJ mol^{-1}) used in this work, shown with those derived from this work using the 0 K onset energies, E_0

	$\Delta_f H^{\theta}_{0K}$	$\Delta_f H^{\theta}_{298K}{}^g$	$H^{\theta}_{298K} - H^{\theta}_{0K}$ (G3B3)*	E_0/eV*	Corresponding process
1,1-C$_2$H$_2$F$_2$	-343.1 ± 2.5^a	-350.2 ± 2.5^a	12.27		
F	77.3 ± 0.3^b	79.4 ± 0.3^b	6.52^b		
CH$_2$=CF$^+$	$976 \pm 9^*$	$973 \pm 9^*$	12.25	14.47 ± 0.1	C$_2$H$_2$F$_2$ → CH$_2$=CF$^+$ + F + e$^-$
C$_2$F$_4$	-669.4 ± 3.3^a	-672.8 ± 3.3^a	16.43		
CF$_3$	-462.8 ± 2.1^c	-465.7 ± 2.1^c	11.55		
CF$_3^+$	413.4 ± 2.0^d	410.2 ± 2.0^d	11.14		
CF	$240.7 \pm 3.9^*$	$243.9 \pm 3.9^*$	8.70	13.717 ± 0.007	C$_2$F$_4$ → CF$_3^+$ + CF + e$^-$
CF$^+$	$1119.1 \pm 4.0^*$	$1122.3 \pm 4.0^*$	8.68	13.740 ± 0.010	C$_2$F$_4$ → CF$^+$ + CF$_3$ + e$^-$
CF$_2$	$-199.7 \pm 5.6^*$	$-199.2 \pm 5.6^*$	10.35	14.16 ± 0.04	C$_2$F$_4$ → CF$_2^+$ + CF$_2$ + e$^-$
CF$_2^+$	-195.0 ± 2.9^e	-194.5 ± 2.9^e	10.35		
CHF$_2^+$	602.4 ± 2.7^f	597.9 ± 2.7^f	10.39	13.856 ± 0.007	C$_2$HF$_3$ → CHF$_2^+$ + CF + e$^-$
C$_2$HF$_3$	$-494.6 \pm 4.8^*$	$-499.9 \pm 4.8^*$	14.31		
CHF$_2$	$-246.7 \pm 4.9^*$	$-250.6 \pm 4.9^*$	10.97	14.16 ± 0.03	C$_2$HF$_3$ → CF$^+$ + CHF$_2$ + e$^-$
CHF$^+$	$1108.0 \pm 5.9^{*,h}$	$1108.1 \pm 5.9^{*,h}$	9.80	14.54 ± 0.02	C$_2$HF$_3$ → CHF$^+$ + CF$_2$ + e$^-$

*This work. ^{a} Feller et al. [56, 62]. ^{b} Chase, JANAF tables [58]. ^{c} Ruscic et al. [57]. ^{d} Bodi et al. [52]. ^{e} Dixon and Feller [60]. ^{f} From Chap. 4, CH$_2$F$_2$, the E_0 of the reaction CH$_2$F$_2$ → CHF$_2^+$ + H + e$^-$ was found to be 13.060 ± 0.015 eV [63]. ^{g} ($H^{\circ}_{298K} - H^{\circ}_{0K}$) values for C, H$_2$ and F$_2$ are 1.05, 8.47 and 8.82 kJ mol^{-1}, respectively [58]. ^{h} Value determined using $\Delta_f H^{\theta}_{0K}$(CF$_2$) = -195.0 kJ mol^{-1} [60]

5.3.5 Trends and Insights into Bonding

We consider five statistical dissociation channels on the ground electronic state and non-statistical F-loss from excited electronic states in $C_2H_{4-n}F_n^+$. The former channels comprise: (1) C–H bond cleavage, (2) statistical C–F bond cleavage, (3) HF loss by way of a four-membered ring transition state, (4) C–C bond cleavage by way of a three-membered ring transition state, and (5) C=C bond cleavage.

The C–H bond becomes progressively stronger with increasing n. In trifluoroethene, with a G3B3 calculated H-loss onset of 15.48 eV, the C–H bond is already too strong to compete successfully with the other dissociation channels. **The F-loss potential energy well also deepens** in the group with increasing n. In mono- and difluoroethene, statistical F-loss competes effectively, whereas statistical F-loss is at most a minor channel in trifluoroethene, and absent in $C_2F_4^+$. In fact, non-statistical F-loss establishes a second dissociation regime in $C_2F_4^+$, in which only the ground electronic state is inaccessible to the reactive flux.

Four-membered ring CHFC transition states may lead to HF loss in n = 1–3, which is the least endothermic channel, albeit with a decreasing margin as n increases, and is absent in the trifluoroethene breakdown diagram because of the large barrier to forming the transition structure. In short, **the four-membered ring transition structure is destabilised** as n increases. Three-membered ring transition states lead to F/H-migration and subsequent C–C bond breaking. The F-transfer transition state in trifluoroethene is calculated to be 0.19 eV lower in energy than H-transfer, but H-transfer may still compete if the tunnelling through the barrier is sufficiently fast. With increasing F-substitution, **the three-membered ring transition states are found to be stabilised**, and the resulting fragments dominate the low-energy breakdown diagram in n = 3 and 4. In contrast with the four-membered ring HF-loss transition state, F-migration takes place at internal energies below the C–C bond energy. This means that the corresponding dissociative photoionization processes are fast, and their endothermicities can be determined based on the breakdown diagram.

Finally, the C=C bond energy decreases from n = 3 to n = 4 as predicted by the perfluoro effect [47]. C=C bond rupture is a minor channel in monofluoroethene, a significant one in n = 2, hardly observed for n = 3, and becomes one of the major regime-one channels in tetrafluoroethene.

5.3.6 Thermochemistry

For the dissociative photoionization reaction $AB + hv \rightarrow A^+ + B + e^-$, the enthalpy of the unimolecular reaction, $\Delta_r H^0$, and the appearance energy of the daughter ion A^+, E_0, are equivalent only at 0 K and in the absence of a reverse barrier, see Fig. 5.1a;

5.3 Results and Discussion

$$E_0 = \Delta_r H^\circ_{0K} = \sum (\Delta_f H^\circ_{0K})_{products} - \sum (\Delta_f H^\circ_{0K})_{reactants} \quad (5.1)$$

Therefore, using 0 K appearance energies with established enthalpies of formation for neutral parent molecules, neutral fragments and daughter ions, the 0 K enthalpy of formation of the least well-determined species can be obtained.

To convert the enthalpy of formation of a molecular species or ion [AB] between 0 and 298 K we use

$$(\Delta_f H^\circ_{298K} - \Delta_f H^\circ_{0K})_{[AB]} = (H^\circ_{298K} - H^\circ_{0K})_{[AB]} - \sum (H^\circ_{298K} - H^\circ_{0K})_{constituent\ elements} \quad (5.2)$$

where the thermal correction for a non-linear molecule is defined as

$$(H^\circ_{298K} - H^\circ_{0K}) \simeq \frac{5}{2}k_BT + \frac{3}{2}k_BT + \sum_{vib} \frac{h\nu}{\exp\left(\frac{h\nu}{k_BT}\right) - 1} = 4k_BT + \sum_{vib} \frac{h\nu}{\exp\left(\frac{h\nu}{k_BT}\right) - 1} \quad (5.3)$$

The electron value of $(H_{298K} - H_{0K})$ is neglected in Eq. 5.2, i.e. we use the stationary electron (or ion) convention for ions at $T > 0$ K [38].

We cannot deduce any thermochemical values from $C_2H_3F^+$ because of the slow HF loss and insufficient resolution of the H-loss signal due to the broadened parent TOF signal. Values derived from the other three molecules can be found in Table 5.1. In 1,1-difluoroethene, the E_0 value for F loss, 14.47 ± 0.1 eV, and $\Delta_f H^0_{0K}$ (1,1-$C_2H_2F_2$) = -343.1 ± 2.5 kJ mol^{-1} [56], yield $\Delta_f H^0_{0K}$ (CH$_2$=CF$^+$) = 976 ± 9 kJ mol^{-1}, converted to 973 ± 9 kJ mol^{-1} at 298 K. This last value can be compared with the previous room temperature value of 951 kJ mol^{-1} [38].

For tetrafluoroethene, $\Delta_f H^0_{0K}$ (C_2F_4) = -669.4 ± 3.3 kJ mol^{-1} [56], the $C_2F_4 \rightarrow CF_3^+ + CF + e^-$ E_0 value of 13.717 ± 0.007 eV, and $\Delta_f H^0_{0K}$ (CF$_3^+$) = 413.4 ± 2.0 kJ mol^{-1} [52], yield $\Delta_f H^0_{0K}$ (CF) = 240.7 ± 3.9 kJ mol^{-1}. This is an improved value upon that of Asher and Ruscic [51], of 251.0 ± 4.6 kJ mol^{-1}, partly because they used an outdated JANAF value which was 14 kJ mol^{-1} too high and partly because they overestimated the CF$^+$/CF$_3^+$ onset energy difference; thus they underestimated the CF$_3$ ionization energy by 0.04 eV. From the $C_2F_4 \rightarrow CF^+ + CF_3 + e^-$ E_0 value of 13.740 ± 0.010 eV, using $\Delta_f H^0_{0K}$ (CF$_3$) = -462.8 ± 2.1 kJ mol^{-1} [57], we present an improved value of $\Delta_f H^0_{0K}$ (CF$^+$) = 1119.1 ± 4.0 kJ mol^{-1}. This is in close agreement to the Burcat value of 1121.9 ± 0.9 kJ mol^{-1} at 0 K [31], though some distance from the Lias and JANAF values of 1131.0 and 1140.0 ± 0.5 kJ mol^{-1} respectively [38, 58]. Using the $C_2F_4 \rightarrow CF_2^+ + CF_2 + e^-$ E_0 value of 14.16 ± 0.04 eV, the IE of CF$_2$ of 11.362 ± 0.03 eV [43] and $\Delta_f H^0_{0K}$ (C_2F_4) of -669.4 ± 3.3 kJ mol^{-1} [56], we obtain $\Delta_f H^0_{0K}$ (CF$_2$) = -199.7 ± 5.6 kJ mol^{-1}. This value may be compared with previous values -182.5 ± 6.3 kJ mol^{-1} (JANAF) [58], -185.3 ± 4.2 kJ mol^{-1} (Berman) [59], -191.7 ± 1.3 kJ mol^{-1} (Burcat) [31], -195.0 ± 2.9 kJ mol^{-1} (Dixon and Feller) [60] and -205.0 kJ mol^{-1} (Lias) [61].

In Chap. 4, the E_0 value of the fast dissociation $CH_2F_2 \rightarrow CHF_2^+ + H + e^-$ was found to be 13.060 ± 0.015 eV [63]. Thus, using the thermochemical network also described in Chap. 4 together with $\Delta_f H^{\theta}_{0K}$ (H) = 216.0 kJ mol^{-1} [58] and $\Delta_f H^{\theta}_{0K}$ (CH_2F_2) = -442.6 ± 2.0 kJ mol^{-1} [64], $\Delta_f H^{\theta}_{0K}$ (CHF_2^+) = 602.4 ± 2.7 kJ mol^{-1} can be derived. This value was then used, together with the now obtained $\Delta_f H^{\theta}_{0K}$ (CF) = 240.7 ± 3.9 kJ mol^{-1} and the E_0 value for $C_2HF_3 \rightarrow CHF_2^+ + CF + e^-$ of 13.856 ± 0.007 eV in trifluoroethene, to derive $\Delta_f H^{\theta}_{0K}$ (C_2HF_3) = -494.6 ± 4.8 kJ mol^{-1}. Burcat [31] and Lias [38] report -485.5 and -485.7 kJ mol^{-1} respectively, for this quantity. From the $\Delta_f H^{\theta}_{0K}$ (C_2HF_3) and $\Delta_f H^{\theta}_{0K}$ (CF^+) derived herein, and the $C_2HF_3 \rightarrow CF^+ + CHF_2 + e^-$ 0 K appearance energy of 14.16 ± 0.03 eV, we obtain $\Delta_f H^{\theta}_{0K}$ (CHF_2) = -246.7 ± 4.9 kJ mol^{-1}. This value is somewhat less than the values of Lias -233.8 ± 5 kJ mol^{-1} [38], Burcat -235.7 kJ mol^{-1} [31], and a more recent ab initio study of -239.4 ± 2.6 kJ mol^{-1} [64]. From the $C_2HF_3 \rightarrow CHF^+ + CF_2 + e^-$ threshold of 14.54 ± 0.02 eV, and $\Delta_f H^{\theta}_{0K}$ (C_2HF_3) derived in this work, the $\Delta_f H^{\theta}_{0K}$ (CHF^+) was found to be 1108.0 ± 5.9 kJ mol^{-1}. The enthalpies of formation derived in this work (asterisked values in Table 5.1) differ from those given in the journal article where this work has been published, due to the use of the E_0 of the reaction $CH_2F_2 \rightarrow CHF_2^+ + H + e^-$ obtained by modelling the breakdown curves from Chap. 4. The article value used an E_0 obtained by eye only.

5.4 Conclusions

The unimolecular dissociation of energy-selected fluorinated ethene cations have been investigated in the 13–25 eV energy range. Four statistical channels, namely HF loss, F loss, direct cleavage of the C=C double bond as well as cleavage of the C–C bond post F or H migration have been discussed in detail, in addition to the non-statistical F-loss channel.

The studied fluorinated ethenes may be divided into two groups, the 'time bombs' (monofluoroethene and 1,1-difluoroethene) and the 'fast dissociators' (trifluoroethene and tetrafluoroethene). In the time bombs, the least endothermic HF loss channel is blocked by a tight 4-membered ring transition state structure. As a result, the parent ions have long lifetimes in the μs timescale at the onset of dissociative photoionization, succeeded by impulsive loss of HF with about 1 eV kinetic energy release. The latter is due to the large reverse barrier, reproduced well by the RAC-RRKM modelled appearance energies. In tri- and tetrafluoroethene, the two main channels at low energies are the post F/H-transfer C–C bond cleavages, in which the charge stays on either fragment. These processes are found

5.4 Conclusions

to take place without an overall reverse barrier, and by taking into account the competitive shifts in the breakdown curves and deriving accurate 0 K appearance energies, we obtain the ionization energy differences for these fragments directly. This is particularly useful in C_2F_4, where it leads to a new, self-consistent set of thermochemical values for the $CF/CF_3/CF^+/CF_3^+$ system, ($\Delta_f H^\theta_{0K}$ = 240.7 ± 3.9, −462.8 ± 2.1, 1119.1 ± 4.0 and 413.4 ± 2.0 kJ mol^{-1} respectively). The ionization energy of CHF_2 has been re-determined to be 8.81 ± 0.02 eV. The ionization energy of CF_3 has also been determined, and at 9.090 ± 0.015 eV is slightly higher than previous values.

As the C–F bond gets progressively stronger with increasing fluorine substitution, while the F/H-atom migration transition state becomes stabilized, statistical F loss becomes less competitive. There is evidence of a higher energy, non-statistical F-loss channel in all four molecules studied, but it is a dominant and exclusive F-loss channel in $C_2F_4^+$. Thanks to a fortunate partitioning of the dissociative photoionization products, we could construct and model a second, regime-two breakdown diagram, in which a sequential CF_2 loss is also included from the F-loss fragment ion, $C_2F_3^+$. By measuring the product energy distribution of the F-loss daughter, we could establish that only the ground electronic state of $C_2F_4^+$ is inaccessible in the non-statistical F loss channel. Therefore, the dissociating excited state $C_2F_4^{*+}$ ion is long-lived, and that the excess energy is statistically redistributed among the nuclear degrees of freedom.

References

1. Sztáray, B., Bodi, A., & Baer, T. (2010). *Journal of Mass Spectrom, 45*, 1233–1245.
2. Blanksby, S. J., & Ellison, G. B. (2003). *Accounts of Chemical Research, 36*, 255–263.
3. Takeshita, K. (1999). *Theoretical Chemistry Accounts, 101*, 343–351.
4. Jarvis, G. K., Boyle, K. J., Mayhew, C. A., & Tuckett, R. P. (1998). *Journal of Physical Chemistry A, 102*, 3219.
5. Jarvis, G. K., Boyle, K. J., Mayhew, C. A., & Tuckett, R. P. (1998). *Journal of Physical Chemistry A, 102*, 3230–3237.
6. Jennings, K. R. (1970). *Organic Mass Spectrometry, 3*, 85.
7. Stadelmann, J.-P., & Vogt, J. (1980). *International Journal of Mass Spectrometry and Ion Physics, 35*, 83–89.
8. Dannacher, J., Schmelzer, A., Stadelmann, J.-P., & Vogt, J. (1979). *International Journal of Mass Spectrometry and Ion Physics, 31*, 175–186.
9. Güthe, F., Locht, R., Leyh, B., Baumgärtel, H., & Weitzel, K.-M. (1999). *Journal of Physical Chemistry A, 103*, 8404–8412.
10. Güthe, F., Baumgärtel, H., & Weitzel, K.-M. (2001). *Journal of Physical Chemistry A, 105*, 7508–7513.
11. Frenking, G., Koch, W., Schaale, M., & Baumgärtel, H. (1984). *International Journal of Mass Spectrometry and Ion Processes, 61*, 305.
12. Booze, J., Weitzel, K.-M., & Baer, T. (1991). *Journal of Chemical Physics, 94*, 3649–3656.
13. Malow, M., Güthe, F., & Weitzel, K.-M. (1999). *Physical Chemistry Chemical Physics, 1*, 1425.

14. Gridelet, E., Dehareng, D., Locht, R., Lorquet, A. J., Lorquet, J. C., & Leyh, B. (2005). *Journal of Physical Chemistry A, 109*, 8225.
15. Bodi, A., Johnson, M., Gerber, T., Gengeliczki, Z., Sztáray, B., & Baer, T. (2009). *Review of Scientific Instruments, 80*, 034101.
16. Johnson, M., Bodi, A., Schulz, L., & Gerber, T. (2009). *Nuclear Instruments and Methods in Physics Research A, 610*, 597–603.
17. Sztáray, B., Bodi, A., & Baer, T. (2010). *Journal of Mass Spectrom, 45*, 1233–1245.
18. Lifshitz, C. (1982). *Mass Spectrometry Reviews, 1*, 309–348.
19. Weitzel, K.-M., Malow, M., Jarvis, G. K., Baer, T., Song, Y., & Ng, C. Y. (1999). *Journal of Chemical Physics, 111*, 8267–8270.
20. Baer, T., Sztáray, B., Kercher, J. P., Lago, A. F., Bodi, A., Skull, C., et al. (2005). *Physical Chemistry Chemical Physics, 7*, 1507–1513.
21. Bieri, G., Niessen, W. V., Åsbrink, L., & Svensson, A. (1981). *Chemical Physics, 60*, 61–79.
22. Buckley, T. J., Johnson, R. D., Huie, R. E., Zhang, Z., Kuo, S. C., & Klemm, R. B. (1995). *Journal of Physical Chemistry, 99*, 4879.
23. Powis, I., Dutuit, O., Richard-Viard, M., & Guyon, P. M. (1990). *Journal of Chemical Physics, 92*, 1643–1652.
24. Smith, D. M., Tuckett, R. P., Yoxall, K. R., Codling, K., Hatherly, P. A., Aarts, J. F. M., et al. (1994). *Journal of Chemical Physics, 101*, 10559–10575.
25. Baer, T., Guerrero, A., Davalos, J. Z., & Bodi, A. (2011). *Physical Chemistry Chemical Physics, 13*, 17791.
26. Berkowitz, J. (1978). *Journal of Chemical Physics, 69*, 3044–3054.
27. Borkar, S., Sztáray, B., & Bodi, A. (2011). *Physical Chemistry Chemical Physics, 13*, 13009–13020.
28. Smith, D. M., Tuckett, R. P., Yoxall, K. R., Codling, K., & Hatherly, P. A. (1993). *Chemical Physics Letters, 216*, 493–502.
29. Shaw, D. A., & Holland, D. M. P. (2008). *Journal of Physics B: Atomic, Molecular and Optical Physics. 41*, 145103–145113.
30. Nenner, I., Guyon, P. M., Baer, T., & Govers, T. R. (1980). *Journal of Chemical Physics, 72*, 439115–439121.
31. Burcat, A., & Ruscic, B. (September 2005). Third Millennium Ideal Gas and Condensed Phase Thermochemical Database for Combustion with Updates from Active Thermochemical Tables, ANL-05/20 and TAE 960, Technion-IIT; Aerospace Engineering, and Argonne National Laboratory, Chemistry Division.September.ftp:technion.ac.il/pub/supported/aetdd/thermodynamics mirrored at http://garfield.chem.elte.hu/Burcat/burcat.html.
32. Lias, S. G. (2011)."Ionization energy evaluation". In P. J Linstrom & W. G Mallard (Eds.), *NIST chemistry webBook, NIST standard reference database number 69* (Vol. 20899). Gaithersburg: National Institute of Standards and Technology.
33. Cox, J. D., Wagman, D. D., & Medvedev, V. A. (1984). *CODATA key values for thermodynamics; hemisphere publishing corp* (Vol. 1). New York.
34. Roorda, M., Lorquet, A. J., & Lorquet, J. C. (1991). *Journal of Physical Chemistry, 95*, 9118–9121.
35. Momigny, J., & Locht, R. (1993). *Chemical Physics Letters, 211*, 161–165.
36. Bodi, A., Stevens, W. R., & Baer, T. (2011). *Journal of Physical Chemistry A, 115*, 726.
37. Ruscic, B., Pinzon, R. E., Morton, M. L., Laszewski, G. V., Bittner, S., Nijsure, S. G., et al. (2004). Journal of Physical Chemistry A, *108*, 9979–9997.
38. Lias, S. G., Bartmess, J. E., Liebman, J. F., Holmes, J. L., Levin, R. D., & Mallard, W. G. (1988). Journal of Physical and Chemical Reference Data, *17*, 1–861.
39. Dyke, J. M., Lewis, A. E., & Morris, A. (1984). *Journal of Chemical Physics, 80*, 1382–1387.
40. Andrews, L., Dyke, J. M., Jonathan, N., Keddar, N., Morris, A., & Ridha, A. (1984). *Journal of Physical Chemistry, 88*, 2364.
41. Litorja, M., & Ruscic, B. (1998). *Journal of Chemical Physics, 108*, 6748–6755.
42. Irikura, K. K., Hudgens, J. W., & Johnson III, R. D. (1995). Journal of Chemical Physics, *103*, 1303.

References

43. Innocenti, F., Eypper, M., Lee, E. P. F., Stranges, S., Mok, D. K. W., Chau, F., et al. (2008). *Chemistry—A European Journal, 14*, 11452–11460.
44. Aue, D. H., & Bowers, M. T. (1979). Gas phase basicities. In M. T Bowers (Ed.), *Gas phase ion chemistry* (Vol. 2, pp. 1–51), New York: Academic Press.
45. Goto, M., Nakamura, K., Toyoda, H., & Sugai, H. (1994). Japanese Journal of Applied Physics Part 1, *33*, 3602.
46. Dearden, D. V., Hudgens, J. W., Johnson, R. D., Tsai, B. P., & Kafafi, S. A. (1992). *Journal of Physical Chemistry, 96*, 585–594.
47. Brundle, C. R., Robin, M. B., Kuebler, N. A., & Harold, B. (1972). *Journal of American Chemical Society, 94*, 1451–1465.
48. Walters, E. A., Clay, J. T., & Grover, J. R. (2005). *Journal of Physical Chemistry A, 109*, 1541.
49. Jarvis, G. K., & Tuckett, R. P. (1998). *Chemical Physics Letters, 295*, 145.
50. Garcia, G. A., Guyon, P. M., & Powis, I. (2001). *Journal of Physical Chemistry A, 105*, 8296–8301.
51. Asher, R. L., & Ruscic, B. (1996). *Journal of Chemical Physics, 106*, 210–221.
52. Bodi, A., Kvaran, Á., & Sztáray, B. (2011). *Journal of Physical Chemistry A, 115*, 13443–13451.
53. Lifshitz, C., & Long, F. A. (1963). *Journal of Physical Chemistry, 67*, 2463–2468.
54. Bodi, A., Shuman, N. S., & Baer, T. (2009). *Physical Chemistry Chemical Physics, 11*, 11013.
55. Galloy, C., Lecomte, C., & Lorquet, J. C. (1982). *Journal of Chemical Physics, 77*, 4522–4529.
56. Feller, D., Peterson, K. A., & Dixon, D. A. (2011). *Journal of Physical Chemistry A, 115*, 3182.
57. Ruscic, B., Michael, J. V., Redfern, P. C., Curtiss, L. A., & Raghavachari, K. (1998). *Journal of Physical Chemistry A, 102*, 10889.
58. Chase, M. W. (1998). Journal of Physical Chemistry Reference Data Monograph, *9*, 1–1951.
59. Berman, D. W., Bomse, D. W., & Beauchamp, J. L. (1981). *International Journal of Mass Spectrometry and Ion Physics, 39*, 263.
60. Dixon, D. A., & Feller, D. (1998). *Journal of Physical Chemistry A, 102*, 8209–8216.
61. Lias, S. G., Karpas, Z., & Liebman, J. F. (1985). *Journal of American Chemical Society, 107*, 6080.
62. Feller, D., Peterson, K. A., & Dixon, D. A. (2011). *Journal of Physical Chemistry A, 115*, 1440.
63. Harvey, J., Tuckett, R. P., & Bodi, A. (2012). *Journal of Physical Chemistry A, 116*, 9696–9705.
64. Csontos, J., Rolik, Z., Das, S., & Kállay, M. (2010). *Journal of Physical Chemistry A, 114*, 13093.

Chapter 6
Threshold Photoelectron Spectra of Four Fluorinated Ethenes from the Ground Electronic State to Higher Electronic States

6.1 Preamble

The work presented in this chapter has been accepted for publication as a journal article entitled 'Vibrational and electronic excitations in fluorinated ethene cations from the ground up' in 2013 by J. Harvey, P. Hemberger, A. Bodi and R. P. Tuckett, in the Journal of Chemical Physics. 2013, issue 138, pages 124301–124313. The majority of the data collection and analysis was performed by myself; however, the assistance lent by Ms. Nicola Rogers, Drs. Matthew Simpson, Andras Bodi, Melanie Johnson, and Professor Richard Tuckett during beamtime with the collection of the data, and Dr. Patrick Hemberger with data analysis is gratefully acknowledged. I particularly wish to thank Dr. Andras Bodi for his assistance and useful discussions relating to the excited dynamics Sect. 6.2.3. The modelling program used to model the threshold photoelectron spectra was developed by Spangenberg et al. [1].

6.2 Introduction

The dissociation dynamics of the four fluorinated ethenes, $C_2H_3F^+$, 1,1-$C_2H_2F_2^+$, $C_2HF_3^+$ and $C_2F_4^+$ have been investigated and the results presented in Chap. 5. In this chapter, the threshold photoelectron spectra (TPES) of the same four molecules are presented. The TPES can reveal information about the nature of the ground electronic states, the following excited electronic states lying at higher energies, and subsequently the nature of the potential energy surfaces.

The perfluoro effect, i.e. π orbital destabilization with respect to σ orbitals, is observed when substituting hydrogen atoms with fluorine atoms in the series of molecules ranging from ethene to tetrafluoroethene [2, 3]. The earliest comprehensive study of the ionization properties of fluorinated ethene molecules was reported by Sell and Kuppermann, who studied the photoelectron angular distributions in the ground and excited state bands of the HeI photoelectron spectra

(PES) of $C_2H_{4-n}F_n$ ($n = 0$–4) molecules [4]. The HeII PES have been recorded and interpreted, among others, by Bieri et al. with many body Green's function calculations [2]. The Franck–Condon factors for the vibrational progressions in the ground state PES bands of C_2H_3F, 1,1-$C_2H_2F_2$ and C_2HF_3 were calculated by Takeshita, based on Hartree–Fock geometries and force constant matrices [5, 6]. However, he did not attempt to compare the theoretical spectra with experiment. High resolution HeI PES and slightly lower resolution threshold photoelectron spectra (TPES) of C_2H_3F and 1,1-$C_2H_2F_2$ have been reported recently by Locht et al. along with ab initio calculations [7, 8]. The latest HeII photoelectron spectrum of C_2HF_3 was recorded by Bieri et al. [2], but neither a high resolution HeI PES nor a TPES has been reported since. The TPES of C_2F_4, recorded by Jarvis et al. [9], significantly improved upon the resolution of the early work by Sell and Kuppermann [4]. Lately, the HeI PES of C_2F_4 has been studied by Eden et al. [10], with an even higher resolution and signal-to-noise ratio.

The dissociative photoionization dynamics of the four aforementioned fluoroethenes have been studied and presented in Chap. 5 [11]. The first dissociative photoionization channel opens up in a Franck–Condon gap above the electronic ground state $\tilde{X}$. It has been stated in Chap. 5 that F-atom loss is initially a statistical process in three of the four molecular ions, (the exception being tetrafluoroethene), which then turns into a largely non-statistical process at higher energies. This conclusion is based predominantly on the correlation of the F-loss fragment ion signal with features of the TPES, indicating isolated state behaviour, in agreement with previous observations [12–15]. However, it was also found in Chap. 5 that the internal energy distribution of the F-loss daughter ion $C_2F_3^+$ from $C_2F_4^+$ can be modelled assuming a purely statistical dissociation from the $\tilde{A}$ electronic state of the parent ion. This result is contrary to previous reports which invoked impulsive processes [9].

In this chapter, the high resolution TPES of the first photoelectron band of these four molecules, i.e. ionization to the ground electronic states and excited electronic states of the cations, are presented. Excited vibrational states are observed with particular clarity and Franck–Condon simulations are employed to assign vibrational progressions in the ground state. This method has successfully been employed in the study of the photoelectron spectra of small systems e.g. vinyl alcohol [16] and much larger molecules such as ovalene, $C_{32}H_{14}$ [16], as well as for interstellar carbenes [17] and diradicals [18]. The simulations are based upon density functional theory (DFT) geometries and Hessians of the neutral molecule and the cation. This goes beyond the cursory assignment based on vibrational spacings and calculated frequencies. Not only do Franck–Condon factors include symmetry considerations *per se*, they also indicate the relative intensities and the band profile based on the predicted geometry change. This can be vitally important in resolving ambiguities for modes with similar frequencies, or for near degenerate vibrational states. Franck–Condon fits based on DFT force constants are also used to study the parent ion geometries with respect to the in silico geometry optimization results. Of particular interest is whether any loss in planarity of the molecule upon ionization occurs. If the C_S symmetry is conserved and the geometry change is insignificant, only totally symmetric vibrational transitions are allowed in photoionization. However, if the cation becomes non-planar, other vibrational transitions can also gain intensity and become observable. This is indeed the case for $C_2H_4^+$ where the ground state ion tunnels through the barrier of planarity to a torsional (dihedral) angle of 29.2° [19]. In this instance, odd quanta of the non-symmetric twist-assisted mode v_4 are given intensity due to vibronic coupling between the $\tilde{X}$ and $\tilde{A}$ cation electronic states [20, 21].

With increasing F-substitution, the excited electronic states become more indiscernible in the TPES. This spectral congestion may lead one to assume that electronically excited state assignments are fraught with dangers. However, Koopmans' theorem holds and our coupled-cluster assignments agree very well with earlier Hartree–Fock calculations [2]. Two further aspects of the electronically excited states are also touched on. First, we tentatively assign vibrational structures observed in the TPES of excited states. Second, the nature and role of the excited states with regard to the various dissociation pathways has been probed. Specifically, we try to explain the multi-modal mechanism of non-statistical F-loss in the $\tilde{C}$ state band in the monofluoroethene cation observed in a Chap. 5 [11], and aided by quantum chemical calculations, generalize the findings to the remaining members of the fluorinated ethene series.

6.3 Results and Discussion

6.3.1 Ground Electronic State of the Cations

Before the results are presented, a note about Stark shifts is made. High fields have been used throughout the work to extract the electrons from the ionization region (and ions in the coincidence work presented in Chaps. 4 and 5). Under such conditions, field ionization can occur where the molecule is subjected to a Stark effect so high that the potential energy barrier binding the electron with the molecule falls below the energy of the electrons orbital, effectively plucking off the electron to leave the molecular ion. As a result, with high fields, the ionization energy may be recorded at a lower energy than when lower fields are used. Chupka gives an approximate expression for the shift in ionization energy in the diabatic limit,

$$6.1 \, \text{cm}^{-1} \sqrt{F/(V \, \text{cm}^{-1})} \tag{6.1}$$

where F is the electric field [22]. The ground electronic state spectrum of $C_2F_4^+$ was recorded at 20 V cm^{-1} and at 120 V cm^{-1} extraction fields, giving shifts of 27 and 66 cm^{-1} respectively. Based on a previous study of Ar, N_2 and CH_3I using the iPEPICO apparatus [23], the threshold photoelectron peak positions could be expected to be Stark shifted by 5 meV to lower energy when applying the higher field [22]. In the same study [23], field effects were not found to play a role in off-resonance threshold photoionization, which suggests that autoionization can compete effectively with field ionization in the absence of long-lived Rydberg states. In $C_2F_4^+$, it was found that the TPES peak positions did not measurably shift as a function of the field strength, indicating fast autoionization and neutral decay channels for high-n Rydberg states are present. Thus, the constant extraction field of 120 V cm^{-1} does not affect the TPES peak positions significantly in this polyatomic molecule. The error for both the adiabatic (*AIE*) and vertical (*VIE*) ionization energies were determined by taking the half width at half maximum of a Gaussian function fitted to the experimental spectrum.

6.3.1.1 Monofluoroethene

The TPES of C_2H_3F and the simulated stick and convoluted spectra are shown in Fig. 6.1a. The HOMO (highest occupied molecular orbital) of the C_S symmetry neutral is the C=C π bonding orbital $(2a'')^2$, and the cation ground state has the term symbol $\tilde{X}\ ^2A''$. The geometry obtained from FCfit show that planarity is conserved upon ionization to the ground electronic state of the cation, however the C=C bond length increases significantly from 1.320 to 1.409 Å, and the C–F bond length decreases from 1.354 to 1.274 Å. Removing an electron from the HOMO, of bonding character between the carbon atoms, leads to an increase in C=C bond length, whereas the C–F bond length decreases, because the HOMO has anti-bonding character between the carbon and fluorine atoms (see also the bottom schematic structure in Fig. 6.6).

Our ground state TPES is in agreement with the lower resolution TPES recorded by Locht et al. and it agrees very well with a deconvoluted HeI PES (giving a resolution of 8 meV compared with their TPES of 25 meV) of the same authors [7]. The adiabatic ionization energy (*AIE*) is found to be 10.364 ± 0.007 eV and the vertical ionization energy (*VIE*) is 10.556 ± 0.007 eV, both in excellent agreement with previously reported values [7] of 10.363 ± 0.004 and 10.558 ± 0.004 eV, respectively. The ab initio frequencies together with observed frequencies and peak positions are given in Tables 6.1 and 6.2.

The hot band at 10.304 eV most likely corresponds to $v_9'' = 1$, i.e. to the $CHF=CH_2$ wagging mode, calculated to be 484 cm^{-1}, comparing exactly with the experimental value of 0.062 eV or 484 cm^{-1}. Also note that the v_9'' vibrational mode has a' symmetry and is totally symmetric. Similarly to the work of Locht et al. [7], the major progression is identified to be due to the v_4 C=C stretching mode, with up to four quanta observed. The vibrational wavenumbers, symmetries and descriptions of the modes Franck–Condon active upon ionization are given in Fig. 6.3 and peak positions and assignments are shown in Fig. 6.1a. The harmonic

6.3 Results and Discussion

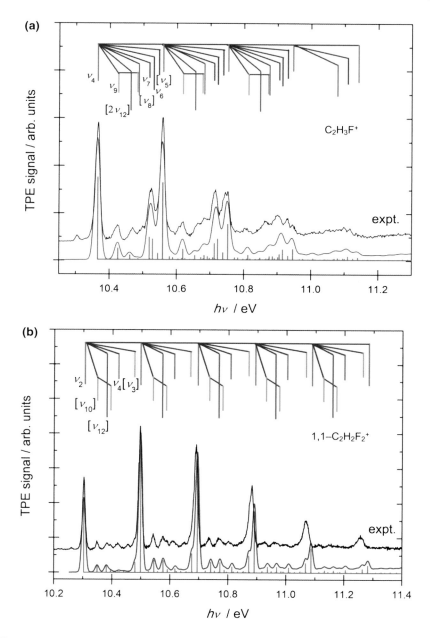

Fig. 6.1 The first TPES band of **a** C_2H_3F and **b** 1,1-$C_2H_2F_2$, is shown with the Franck–Condon stick (*blue stick*) simulations and convoluted curve (*blue curves*). Reassignments of vibrational modes are indicated by square brackets

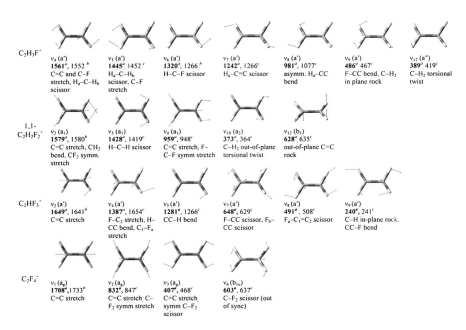

Fig. 6.2 Franck–Condon active vibrational modes of C_2H_3F, 1,1-$C_2H_2F_2$, C_2HF_3 and C_2F_4 upon ionization. [a]B3LYP/6-311 ++G(d,p) harmonic frequencies. [b]Harmonic frequencies derived by Morse-fitting of the vibrational progressions (see text). [c]Frequencies corresponding to the $1 \leftarrow 0$ transition as observed in the TPES

frequency (which is not measured directly), for this mode is determined by fitting the vibrational transitions ($v_4 = 0$–4) to a Morse potential thereby accounting for the anharmonicity, using the well-known [24] approximation

$$E(v+1) - E(v) = hv_0 - \left[(v+1)(hv_0)^2/2D_e\right] \quad (6.2)$$

where hv_0 is the harmonic vibrational frequency and D_e is the dissociation energy [11]. The v_4 harmonic frequency of 1552 cm^{-1} (Fig. 6.3) is in excellent agreement with the B3LYP prediction of 1561 cm^{-1}. Discrepancies in the energies between the simulated and the experimental spectra towards higher eV are due to anharmonicity, which is disregarded in the harmonic model of FCfit. Using the relationship $x_e = hv/4D_e$, the anharmonicity constant, x_e, is determined to be 0.00514. D_e is taken from Chap. 5.

Aside from the v_4 progression, some of the assignments of the remaining weak and complex progressions differ from those of Locht et al. The assignment of the first peak to high energy of the origin band at 10.422 eV to be v_9 of a' symmetry is in agreement with that proposed by Locht et al. However, the second peak at 10.468 eV is 0.102 eV (823 cm^{-1}) higher than the origin band, whilst the next member in the progression at 10.662 eV has a difference of 0.106 eV (855 cm^{-1})

6.3 Results and Discussion

Table 6.1 Calculated (B3LYP) and experimental vibrational modes, symmetries and wavenumbers ($\tilde{v}$) of C_2H_3F $\tilde{X}^1A'$ and $C_2H_3F^+$ $\tilde{X}^2A''$

Symmetry	$\tilde{v}$ C_2H_3F $\tilde{X}^1A'$ (cm^{-1})	$\tilde{v}$ $C_2H_3F^+\tilde{X}^2A''$ (cm^{-1})	Expt (cm^{-1})	Description
v_1 (a')	3257	3252		H_a–C–H_b asymmetric stretch, H–C=C bend
v_2 (a')	3204	3170		C–H stretch, H_a–C stretch
v_3 (a')	3162	3129		H_a–C–H_b symmetric stretch, C–H stretch
v_4 (a')	1703	**1561**[a]	1552[b]	C=C stretch, F–C–H and H_a–C–H_b scissor
v_5 (a')	1407	**1445**[a]	1452	H_a–C–H_b scissor, C–F stretch
v_6 (a')	1329	**1320**[a]	1266	H–C–F scissor
v_7 (a')	1153	**1242**[a]	1266	H_a–C=C bend, C–F stretch
v_8 (a')	929	**981**[a]	1077	Asymmetric H_a–C–C bend
v_9 (a')	**484**[a]	**486**[a]	467	F–CC bend, C–H_2 in plane rock
v_{10} (a'')	961	1016		Out-of-plane CH_2=CHF rock
v_{11} (a'')	892	871		Out-of-plane asymmetric CH_2=CHF rock
v_{12} (a'')	726	**389**[a]	419	C–H_2 torsional twist

Figures in **bold** indicate that [a] Franck–Condon activity is observed in these vibrations in the first photoelectron band
[b] Harmonic frequency derived by Morse fitting of the vibrational progression (see text)

from the 1v_4 peak. The average of the two values is 839 cm^{-1}. This peak is assigned to two quanta of the a'' symmetry v_{12} mode, calculated at 389 cm^{-1}. Indeed, the intensities of 2v_{12} are well reproduced in the Franck–Condon simulation. Note that even-quanta transitions of non-totally symmetric modes are allowed, as a'' × a'' = a'. Locht et al. assign this progression as one quantum of the v_8 mode [7]. This mode has the correct a' symmetry to be observed in odd quanta, but its calculated value at 981 cm^{-1} is significantly higher than the measured 839 cm^{-1} level spacing. The Franck–Condon simulation places this mode at a somewhat higher energy. All peaks previously attributed by Locht et al. as ($nv_4 + v_8$) are re-assigned to ($nv_4 + 2v_{12}$).

The next nearest peak towards the $v_4 = 1$ transition at 10.523 eV has been assigned by Locht et al. to the v_7 mode, a H_a–C=C scissor (where H_a is the hydrogen *cis* to the fluorine, see Table 6.3). The Franck–Condon simulation indicates that both v_6, a H–C–F scissor and v_7 contribute to the peak in the experimental spectrum. B3LYP calculates both vibrations to have a' symmetry with vibrational wavenumbers of 1320 and 1242 cm^{-1}, respectively, to be compared with our experimental value of 0.157 eV or 1266 cm^{-1}. Comparison of the stick and the convoluted spectra suggests that v_6 and v_7 are indeed both blended in the peak at 10.523 eV. There was some ambiguity over the assignment of a weak peak at 10.498 eV [7]. It is comprised of two modes, v_8 (a') with 2v_9 (a'), which are only 9 cm^{-1} apart with comparable Franck–Condon factors. In this instance, it can be said with a degree of certainty that the assignment is not simply a matter of either v_8 or 2v_9, but both transitions are in fact present.

Table 6.2 Assignment of peaks in the first photoelectron band of C_2H_3F

Energy (eV)	Assignment
10.304	Hot band $v_9'' = 1$
10.364	0–0
10.422	v_9
10.468	$2v_{12}$
10.498	$2v_9$ and v_8 together
10.523	v_6 and v_7 together
10.544	v_5
10.556	v_4
10.608	?
10.614	$v_4 + v_9$
10.662	$v_4 + 2v_{12}$
10.680	$v_4 + v_8$, $v_4 + 2v_9$
10.714	$v_4 + v_6$, $v_4 + v_7$
10.737	$v_4 + v_5$
10.746	$2v_4$
10.806	$2v_4 + v_9$
10.837	$2v_4 + 2v_{12}$
10.861	$2v_4 + v_8$, $2v_4 + 2v_9$
10.900	$2v_4 + v_7$
10.928	$2v_4 + v_5$
10.928	$3v_4$
11.096	$3v_4 + v_6$
11.116	$4v_4$

6.3.1.2 1,1-Difluoroethene

Figure 6.1b shows the TPES of 1,1-$C_2H_2F_2$ together with the simulated stick and convoluted spectra. The $(2b_1)^2$ HOMO of the C_{2v} neutral means the cation ground state is $\tilde{X}\ ^2B_1$. Ab initio calculations show there is a significant increase in the C=C bond length from 1.317 to 1.412 Å upon ionization, and a smaller decrease in the C–F bond length from 1.327 to 1.264 Å. The FCfit analysis results in a small twisting of the CF_2 group with respect to the CH_2 group upon ionization (i.e. the dihedral angle of F–C=C–H changes from 180 to 177°), so the planarity of the molecule is lost in the ground state of the cation. For clarity, C_{2v} notation is retained the for the vibrational mode symmetries in the cation. Our spectrum agrees well with both the TPES recorded at lower resolution and the HeI PES recorded at a comparable resolution by Locht et al. [8]. The *AIE* is 10.303 ± 0.005 eV, and the *VIE* is 10.496 ± 0.005 eV. The major vibrational progression has been assigned to the nv_2 ($n = 0$–6) C=C stretching mode of a_1 symmetry and the peak positions given in Table 6.4 are in excellent agreement with those of Locht et al. [8]. Our value for the harmonic frequency v_2 of 1580 cm^{-1}, obtained from Morse fitting of the progression, is in stunning agreement with the ab initio value of 1579 cm^{-1}. The anharmonicity constant, x_e, is

6.3 Results and Discussion

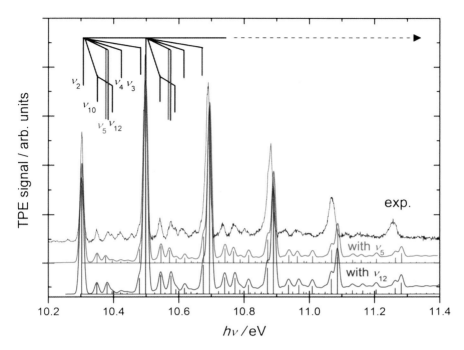

Fig. 6.3 The first TPES band of 1,1-$C_2H_2F_2$ is shown with the Franck–Condon fits including the v_5 mode of (a_2) symmetry with a calculated B3LYP frequency of 583 cm^{-1} (*red line*) v_{12} mode of (b_1) symmetry with a calculated B3LYP frequency of 628 cm^{-1} (*blue line*)

determined to be 0.0046. Three further minor progressions are also identified and their peak positions are in reasonable agreement with the HeI study of Locht et al. [8]. Vibrational assignments are also given in Fig. 6.1b. For the sake of clarity, the ab initio frequencies, observed frequencies and peak positions are given in Tables 6.4 and 6.5.

There are several minor peaks sandwiched in between the v_2 peaks of the main progression. The first one is observed at 10.348 eV and is best assigned to one quantum of v_{10} (a_2) by FCfit. The experimental value of $v_{10} = 363$ cm^{-1} is in agreement with the ab initio result of 373 cm^{-1}. Locht et al. assign this peak to v_9 (b_2) [8], for which the calculated value of 417 cm^{-1} is much higher than the experimental value. It applies to both assignments, however, that these non-totally symmetric vibrations should be forbidden. However, v_{10} is seen, albeit weakly. A possible explanation of how v_{10} is observed could be linked to the loss of planar symmetry upon ionization.

Vibronic coupling between the $\widetilde{X}$ $^2B_1$ and $\widetilde{A}$ $^2B_2$ states of 1,1-$C_2H_2F_2^+$ which is mediated by the v_{10} (a_2) twisting vibrational mode can occur, suggesting a conical intersection is at play, according to the symmetry requirement [25];

Table 6.3 Geometries of the ground electronic states of the neutral and cation of fluorinated ethenes

		Neutral	Cation	Cation
		B3LYP/6-311 ++G(d,p)		FCFIT
C_2H_3F		$\widetilde{X}^1A'$	$\widetilde{X}^2A''$	$\widetilde{X}^2A''$
	C_1–H [Å]	1.083	1.090	1.090
	C_1–F [Å]	1.354	1.274	1.280
	C_1–C_2 [Å]	1.320	1.409	1.404
	C_2–H_a [Å]	1.084	1.088	1.087
	C_2–H_b [Å]	1.081	1.086	1.086
	H_b–C_2–H_a [°]	119.0	120.4	120.1
	H_a–C_2–C_1 [°]	121.7	120.2	120.5
	C_2–C_1–F [°]	122.0	119.0	119.0
	F–C_1–H [°]	111.6	115.3	115.7
	H–C_2–C_1–H_b [°]	180	180	180
1,1–$C_2H_2F_2$		$\widetilde{X}^1A_1$	$\widetilde{X}^2B_1$	$\widetilde{X}^2B_1$
	C_1–H [Å]	1.079	1.085	1.085
	C_1–F [Å]	1.327	1.264	1.263
	C_1–C_2 [Å]	1.317	1.412	1.404
	H–C_2–H [°]	120.4	121.9	121.7
	H–C_2–C_1 [°]	119.8	119.1	119.1
	C_2–C_1–F [°]	125.3	122.4	122.5
	F–C_1–F [°]	109.4	115.2	115.0
	H–C_2–C_1–F [°]	180.0	180.0	177.0
C_2HF_3		$\widetilde{X}^1A'$	$\widetilde{X}^2A''$	$\widetilde{X}^2A''$
	C_2–H [Å]	1.079	1.088	1.088
	C_2–F [Å]	1.345	1.275	1.277
	C_1–C_2 [Å]	1.323	1.418	1.414
	C_1–F_a [Å]	1.325	1.265	1.266
	C_1–F_b [Å]	1.318	1.262	1.264
	H–C_2–C_1 [°]	123.4	123.0	123.3
	F–C_2–H [°]	116.0	119.1	118.9
	C_2–C_1–F_a [°]	122.9	120.5	120.5
	F_a–C_1–F_b [°]	111.8	117.1	117.0
	H–C_2–C_1–F_a [°]	180	180	180
C_2F_4		$\widetilde{X}^1A_g$	$\widetilde{X}^2B_{3u}$	$\widetilde{X}^2B_{3u}$
	C_1–F [Å]	1.321	1.265	1.262
	C_1–C_2 [Å]	1.322	1.418	1.427
	F–C_1–C_2 [°]	123.4	120.8	120.7
	F–C_1–F [°]	113.1	118.3	118.7
	F–C_2–C_1–F [°]	180	180	180

6.3 Results and Discussion

Table 6.4 Calculated (B3LYP) and experimental vibrational modes, symmetries and wavenumbers ($\tilde{v}$) of 1,1-$C_2H_2F_2$ $\tilde{X}^1A_1$ and 1,1-$C_2H_2F_2^+$ $\tilde{X}^2B_1$

Symmetry	$\tilde{v}$ 1,1-$C_2H_2F_2\tilde{X}^1A_1$ (cm^{-1})	$\tilde{v}$ 1,1-$C_2H_2F_2^+\tilde{X}^2B_1$ (cm^{-1})	Expt (cm^{-1})	Description
v_1 (a_1)	3191	3139		C–H symmetric stretch
v_2 (a_1)	1769	1579[a]	1580[b]	C=C stretch, F–C–F symmetric stretch
v_3 (a_1)	1407	1428[a]	1419	H–C–H symmetric scissor
v_4 (a_1)	927	959[a]	948	C=C stretch, F–C–F symmetric stretch
v_5 (a_1)	546	583		C–F_2 scissor
v_6 (b_2)	3292	3268		C–H asymmetric stretch
v_7 (b_2)	1280	1515		F–C–F asymmetric stretch, C–H_2 bend
v_8 (b_2)	955	1006		H–CC asymmetric bend
v_9 (b_2)	438	417		F–CC asymmetric bend
v_{10} (a_2)	718	373[a]	364	C–H_2 out-of-plane torsional twist
v_{11} (b_1)	834	924		Out-of-plane H–C–H rock
v_{12} (b_1)	624	628[a]	635	Out-of-plane C=C rock

[a] Franck–Condon activity is observed in these vibrations in the first photoelectron band
[b] Harmonic frequency derived by Morse fitting of the vibrational progression (see text)

$$\Gamma_e^{\tilde{X}} \otimes \Gamma_e^{\tilde{A}} \supset \Gamma_v \quad (6.3)$$

This is satisfied with the coupling of the vibronic symmetry of the $\tilde{X}$ and $\tilde{A}$ states, $^2B_1 \otimes {}^2B_2 = A_2$, giving the symmetry of the vibrational mode, v_{10}. Therefore, the $^2B_1$ and $^2B_2$ ion states are coupled when the molecule is twisted, and intensity is given to the v_{10} mode upon ionization. As the predicted torsional angle is only ±3°, the double-minimum potential energy curve must be very shallow with the $v_{10} = 0$ and 1 levels most likely above the barrier. In contrast, the non-adiabatic coupling (via a conical intersection) between the $\tilde{X}$ and $\tilde{A}$ states of $C_2H_4^+$, which is mediated by the torsional mode, produces a torsional angle at the minima of the ground state which is much larger, ±29°, with a barrier height of 357 cm^{-1} [19]. The difference in the extent of coupling and therefore the amount of twist seen could be due, in part, to the larger difference in energy [26] between the $\tilde{X}$ and $\tilde{A}$ states of 1,1-$C_2H_2F_2^+$ of 4.31 eV compared with that of $C_2H_4^+$ of 2.31 eV [4].

The second of these minor peaks at 10.382 eV lies 637 cm^{-1} above the band origin and is also well reproduced in the Franck–Condon fitting by the v_{12} mode of b_1 symmetry at 628 cm^{-1}. This is the only mode within 50 cm^{-1} of the experimental value of 635 cm^{-1}. Populating the F–C–F scissor v_5 mode of a_1 symmetry, with a frequency of 583 cm^{-1} gives a similar fit but it is slightly lower in energy than the v_{12} mode (see Fig. 6.4). The Franck–Condon simulation yields a third minor peak at 10.397 eV, which corresponds to a shoulder in the experimental spectrum at 10.394 eV, assigned as two quanta in the v_{10} mode of a_2 symmetry

Table 6.5 Assignment of peaks in the first photoelectron band of 1,1-$C_2H_2F_2$

Energy (eV)	Assignment
10.260	Hot band
10.303	0–0
10.348	v_{10}
10.382	v_{12}
10.394	$2v_{10}$
10.420	v_4
10.456	
10.479	v_3
10.496	v_2
10.542	$v_2 + v_{10}$
10.577	$v_2 + v_{12}$
10.590	$v_2 + 2v_{10}$
10.610	$v_2 + v_4$
10.651	
10.675	$v_2 + v_3$
10.690	$2v_2$
10.734	$2v_2 + v_{10}$
10.768	$2v_2 + v_{12}$
10.779	$2v_2 + 2v_{10}$
10.803	$2v_2 + v_4$
10.847	$2v_2 + v_3$
10.882	$3v_2$
10.925	$3v_2 + v_{10}$
10.958	$3v_2 + v_{12}$
10.975	$3v_2 + 2v_{10}$
11.000	$3v_2 + v_4$
11.034	$3v_2 + v_3$
11.069	$4v_2$
11.111	$4v_2 + v_{10}$
11.148	$4v_2 + v_{12}$
11.257	$5v_2$

(2×373 cm^{-1}). The difference between the band origin and this shoulder is 0.091 eV or 734 cm^{-1} which is close to the calculated value of 746 cm^{-1}. This mode becomes allowed under symmetry considerations even without the breakdown of planarity. Locht et al. do not resolve this doublet and assign the single peak as $2v_9$ [8]. Even if the HeI peak at 10.347 ± 0.004 eV had been correctly assigned as v_9 by Locht et al., the 0.042 eV or 338 cm^{-1} spacing to the 10.389 ± 0.005 eV peak would still mean the latter is unlikely to be $2v_9$ with $v_9 = 417$ cm^{-1}. Note that both the v_{10} and v_{12} modes involve a twisting of the CH$_2$ moiety which changes the dihedral angle, whereas the v_9 mode consists of a F–CC asymmetric in-plane bend (wagging motion) between the CH$_2$ and CF$_2$ moieties which does not cause a change in this angle. When populating the v_9 mode instead of either v_{10} or v_{12}, the resulting spectrum is not a satisfactory fit to the

6.3 Results and Discussion

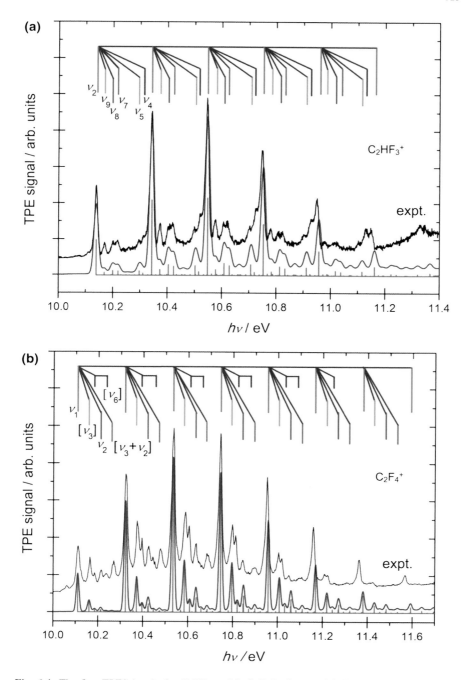

Fig. 6.4 The first TPES band of **a** C_2HF_3 and **b** C_2F_4 is shown with the Franck–Condon stick (*blue stick*) simulations and convoluted curve (*blue curves*). Reassignments of vibrational modes are indicated by square brackets

experimental spectra as the H–C=C angle of the cation becomes drastically reduced. In addition, the experimental spectrum cannot be faithfully reproduced when a planar cation geometry is retained, confirming that the twist gives rise to the observation of both the v_{10} and v_{12} modes.

The fourth peak to high energy of the origin band is at 10.420 eV. This is assigned to v_4 (a_1) = 1, with this F–C–F symmetric stretching mode at 948 cm^{-1}, to be compared with the ab initio value of 959 cm^{-1}. Finally, there is a partially resolved shoulder (starting at 10.479 eV) to lower energy of each peak of the main v_2 (a_1) progression. Based on the Franck–Condon simulation, this may correspond to the v_3 (a_1) vibrational mode, observed experimentally at around 1418 cm^{-1}, which can be compared with the ab initio value of 1428 cm^{-1}. This progression was not observed by Locht et al. [8], although the peaks in their v_2 progression do appear to be slightly asymmetric. The same pattern of peaks due to the vibrational modes v_{10}, v_{12}, $2v_{10}$, v_4 and v_3 is repeated for members of the main progression of nv_2 ($n = 0$–3).

6.3.1.3 Trifluoroethene

Figure 6.5a shows the TPES of C_2HF_3, the simulated stick and convoluted spectra together with the vibrational assignments. Although some vibrational structure has been observed in the ground-state PE band by others [2, 4], this is the first high resolution TPES of this molecule reported in the literature. The $(4a'')^2$ HOMO of the neutral C_S C_2HF_3 molecule has C=C π orbital character, and the cation electronic ground state has the term symbol $\tilde{X}\,^2A''$. Similarly with the previous fluorinated ethenes, ab initio calculations show an increase in the C=C bond length from 1.323 to 1.418 Å consistent with removing an electron from the C=C π orbital, and a decrease in all C–F bond lengths of ≈ 0.06 Å. The geometry obtained from FCfit shows planarity is retained within the ion.

The first photoelectron band, corresponding to the $\tilde{X}\,^2A''$ ground state of $C_2HF_3^+$ is comprised of a series of sharp and well defined peaks. The *AIE* and *VIE* are 10.138 ± 0.007 and 10.544 ± 0.007 eV, respectively. Previous literature values are scarce with the notable exception of the work by Bieri et al., who reported the *AIE* as 10.14 eV and the *VIE* as 10.62 eV [2], both in close agreement with the values of this work. The band is dominated by a vibrational progression of nv_2 mode ($n = 0$–5) which corresponds to the C=C stretching mode. The Morse-fitted vibrational harmonic frequency of this band is determined to be 1641 cm^{-1}, which is in close agreement with the ab initio value of 1649 cm^{-1}.

The anharmonicity constant, x_e, is determined to be 0.000781. This assignment is in agreement with the early angle-resolved photoelectron spectrum of Sell and Kuppermann [4] (Tables 6.6 and 6.7). There are five other, less intense vibrational progressions amidst members of the v_2 progression. With the aid of FCfit, they are assigned to v_9 (C–H in plane rock and CC–F bend) at 241 cm^{-1}, v_8 (F_a–C_1C_2 scissor, where F_a is *cis* to the hydrogen) at 508 cm^{-1}, v_7 (F–CC scissor and F_b–CC scissor

6.3 Results and Discussion

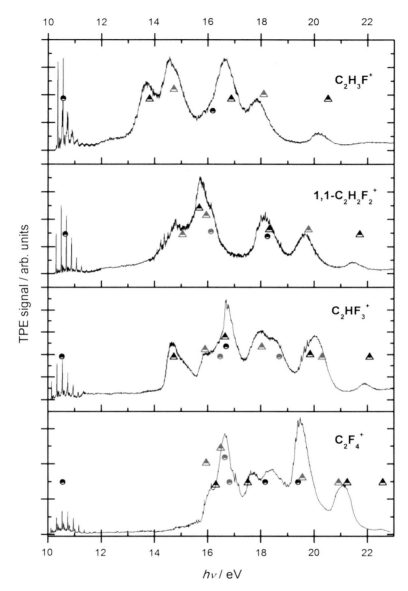

Fig. 6.5 Complete valence threshold photoelectron spectra of C_2H_3F, $1,1\text{-}C_2H_2F_2$, C_2HF_3 and C_2F_4. The EOM-IP-CCSD/cc-pVTZ computed vertical ionization energies are shown by the symbols. Different symbols represent different ion states according to their (approximate, see text) C_{2v} character: A_1 (*black triangle*), A_2 (*light circle*), B_1 (*black circle*), B_2 (*light triangle*)

where F_b is *trans* to the hydrogen) at 629 cm^{-1}, v_5 (C–H wag) at 1266 cm^{-1}, and v_4 (F–C$_2$ stretch, H–CC bend and C$_1$–F$_a$ stretch) at 1654 cm^{-1}. The overall agreement between experiment and fit is excellent. Figure 6.3 shows the calculated and

Table 6.6 Calculated (B3LYP) and experimental vibrational modes, symmetries and wavenumbers ($\tilde{v}$) of C_2HF_3 $\tilde{X}^1A'$ and $C_2HF_3^+$ $\tilde{X}\,^2A''$

Symmetry	$\tilde{v}$ C_2HF_3 $\tilde{X}^1A'$ (cm^{-1})	$\tilde{v}$ $C_2HF_3^+$ $\tilde{X}^2A''$ (cm^{-1})	Expt (cm^{-1})	Description
v_1 (a')	3252	3186		C–H stretch
v_2 (a')	1824	**1649**[a]	1641[b]	C=C stretch
v_3 (a')	1352	1535		F_a–C_1–F_b asymmetric stretch, C–H wag
v_4 (a')	1251	**1387**[a]	1654	C–H wag, F_a–C_1–F_b asymmetric stretch,
v_5 (a')	1160	**1281**[a]	1266	F–C_2 stretch
v_6 (a')	933	952		C–H wag, F_a–C_1–F_b symmetric stretch
v_7 (a')	621	**648**[a]	629	F–CC scissor, F_b–CC scissor
v_8 (a')	483	**491**[a]	508	F_a–C_1–C_2 scissor
v_9 (a')	233	**240**[a]	241	C–H in plane rock, CF_2 bend
v_{10} (a'')	782	812		Out-of-plane C–H rock
v_{11} (a'')	575	578		Out-of-plane C–C–H rock
v_{12} (a'')	306	223		Out-of-plane H rock

[a] Franck–Condon activity is observed in these vibrations in the first photoelectron band. [b] Harmonic frequency derived by Morse fitting of the vibrational progression (see text)

experimental vibrational modes which are active upon ionization and their symmetries. All six active modes are of a' symmetry and satisfy selection rules. Furthermore, just as for monofluoroethene with C_S symmetry, no single quantum of a'' vibrational modes are observed, consistent with a planar cation. Unlike in monofluoroethene, double quanta of a'' modes are not observed in trifluoroethene either. The same pattern of peaks due to the vibrational modes v_9, v_8, v_7, v_5 and v_4 is observed toward higher energy from the main progression peaks nv_2 ($n = 0$–4).

6.3.1.4 Tetrafluoroethene

C_2F_4 has the highest symmetry of the four molecules studied, belonging to the D_{2h} point group. The HOMO of the neutral is the C=C π bonding orbital, $(2b_{3u})^2$, and the cation ground state has the term symbol $\tilde{X}\,^2B_{3u}$ [2]. Using FCfit, the ground state geometry of the cation is confirmed to be planar, and only totally symmetric vibrations in the D_{2h} point group should be observed in the photoelectron spectrum with the highest intensity. As previously, there is an increase in the C=C bond length of 0.096 Å, a decrease in the C–F bond length of 0.056 Å. Overall across the series, the increase in C=C bond length upon ionization becomes greater with increasing F-substitution, but the decrease in C–F bond length is reduced, in accordance with the perfluoro effect, i.e. σ molecular orbitals are strongly stabilized by mixing of the ethylene group orbitals with the electronegative F-atom σ orbitals. By contrast, the mixing and stabilization of the π orbitals is much smaller and so strong C–F π anti-bonding character dominates [3].

6.3 Results and Discussion

Table 6.7 Assignment of peaks in the first photoelectron band of C_2HF_3

Energy (eV)	Assignment
10.073	Hot band
10.097	Hot band
10.108	Hot/sequence band
10.138	0–0
10.168	v_9
10.201	v_8
10.216	v_7
10.295	v_5
10.331	v_4
10.342	v_2
10.373	$v_2 + v_9$
10.407	$v_2 + v_8$
10.418	$v_2 + v_7$
10.499	$v_2 + v_5$
10.519	$v_2 + v_4$
10.544	$2v_2$
10.576	$2v_2 + v_9$
10.605	$2v_2 + v_8$
10.619	$2v_2 + v_7$
10.702	$2v_2 + v_5$
10.726	$2v_2 + v_4$
10.746	$3v_2$
10.778	$3v_2 + v_9$
10.809	$3v_2 + v_8$
10.822	$3v_2 + v_7$
10.901	$3v_2 + v_5$
10.927	$3v_2 + v_4$
10.949	$4v_2$
10.978	$4v_2 + v_9$
11.011	$4v_2 + v_8$
11.021	$4v_2 + v_7$
11.101	$4v_2 + v_5$
11.127	$4v_2 + v_4$
11.146	$5v_2$

The first photoelectron band seen in Fig. 6.5b is assigned to the ground state of $C_2F_4^+$, $\tilde{X}^2B_{3u}$. It is composed of several well-defined vibrational progressions, the most prominent being the nv_1 ($n = 0$–7), the C=C stretching mode at 1708 cm^{-1}, in good agreement with the experimental value from the Morse-fitted progression of 1733 cm^{-1}. The anharmonicity constant, x_e, is determined to be 0.00366. The calculated and experimental frequencies are given in Fig. 6.3. The *AIE* and *VIE* are 10.110 ± 0.009 and 10.535 ± 0.009 eV, respectively. Five additional but less intense vibrational progressions are in between the members of the v_1 progression. Three have been assigned as v_3, v_2 and ($v_3 + v_2$), with experimental frequencies of

Table 6.8 Calculated (B3LYP) vibrational modes, symmetries and wavenumbers ($\tilde{v}$) of $C_2F_4 \tilde{X}^1A_g$ and $C_2F_4^+ \tilde{X}^2B_{3u}$

Symmetry†	$\tilde{v}$ C_2F_4 $\tilde{X}^1A_g$ (cm^{-1})	$\tilde{v}$ $C_2F_4^+$ $\tilde{X}^2B_{3u}$ (cm^{-1})	Expt Value (cm^{-1})	Description
v_1 (a$_g$)[a]	1908	**1708**[b]	1733[c]	C=C stretch
v_2 (a$_g$)[a]	788	**832**[b]	847	C=C stretch, symmetric C–F$_2$ stretch
v_3 (a$_g$)	397	**407**[b]	468	C=C stretch, symmetric C–F$_2$ scissor
v_4 (a$_u$)	198	140		CF$_2$ torsional twist
v_5 (b$_{1u}$)	1173	1275		Symmetric C–F$_2$ stretch, (out of synchronicity)
v_6 (b$_{1u}$)	551	**603**[b]	637	CC–F bend (out of synchronicity)
v_7 (b$_{2g}$)	556	599		Out-of-plane C=C umbrella model
v_8 (b$_{2u}$)	1312	1536		Asymmetric C–F$_2$ stretch (in synchronicity)
v_9 (b$_{2u}$)	211	215		CC–F bend
v_{10} (b$_{3g}$)	1310	1532		Asymmetric C–F$_2$ stretch, (out of synchronicity)
v_{11} (b$_{3g}$)	550	536		CC–F bend
v_{12} (b$_{3u}$)	414	438		Out-of-plane C=C umbrella mode

[a] The labelling of v_1 and v_2 is reversed from the nomenclature used by Jarvis et al. [9]
[b] Franck–Condon activity is observed in these vibrations in the first photoelectron band
[c] Harmonic frequency derived by Morse fitting of the vibrational progression (see text)
† Note the Mulliken convention is used throughout this work, where the *z-axis* is along the C=C bond and the *x-axis* is perpendicular to the molecular plane. The Herzberg convention as used by Eden et al. [8] has the *x-axis* along the C=C bond and the *z-axis* perpendicular to the molecular plane

468, 847 and 1315 cm^{-1} respectively (see Fig. 6.5b) and, as expected, all identified modes are of a$_g$ symmetry (Tables 6.8 and 6.9).

In an earlier lower resolution TPE study by Jarvis et al. [9], only the v_1, v_2 and v_3 modes were observed (note that the numbering of v_1 and v_2 is reversed in both the Jarvis et al. and Brundle et al. studies) [3]. Following a subtraction procedure to allow for the effects of second-order harmonic radiation delivered from the grating monochromator at the beamline, Jarvis et al. determined the *AIE* to be 10.0 ± 0.1 eV, and the vibrational frequencies of these three modes to be 1686, 766 and 371 cm^{-1} [9], in reasonable agreement with those determined from the present work of 1733, 847 and 468 cm^{-1}, respectively. The ground state of $C_2F_4^+$ was also studied by HeI photoelectron spectroscopy at a resolution of 0.022 eV by Eden et al. [10]. The first and third progressions were also identified by Eden et al. as v_1 (C=C stretch) and v_2 (C=C stretch and C–F$_2$ symmetric stretch) [10]. However, there is disagreement between the assignment of the second and fourth progression, which they assign by comparison with the infrared spectrum of neutral C_2F_4, as the v_6 mode of b$_{1g}$ symmetry and the v_{11} mode of b$_{3u}$ symmetry using the Herzberg convention [27]. In the Mulliken convention used here, these vibrations are the v_{11} (b$_{3g}$) CC–F bend and v_5 (b$_{1u}$) symmetric C–F$_2$ stretch modes [28]. Since there is no change in the molecular symmetry, odd-quantum transitions

Table 6.9 Assignment of peaks in the first photoelectron band of C_2F_4

Energy (eV)	Assignment
10.504	Hot/sequence band
10.110	0–0
10.168	v_3
10.188	v_6
10.215	v_2
10.235	$2v_6$
10.273	$v_3 + v_2$
10.326	v_1
10.378	$v_1 + v_3$
10.396	$v_1 + v_6$
10.427	$v_1 + v_2$
10.451	$v_1 + 2v_6$
10.482	$v_1 + v_3 + v_2$
10.535	$2v_1$
10.592	$2v_1 + v_3$
10.606	$2v_1 + v_6$
10.639	$2v_1 + v_2$
10.658	$2v_1 + 2v_6$
10.695	$2v_1 + v_3 + v_2$
10.746	$3v_1$
10.800	$3v_1 + v_3$
10.815	$3v_1 + v_6$
10.847	$3v_1 + v_2$
10.872	$3v_1 + 2v_6$
10.899	$3v_1 + v_3 + v_2$
10.954	$4v_1$
11.007	$4v_1 + v_3$
11.016	$4v_1 + v_6$
11.054	$4v_1 + v_2$
11.095	$4v_1 + v_3 + v_2$
11.159	$5v_1$
11.223	$5v_1 + v_3$
11.273	$5v_1 + v_2$
11.364	$6v_1$
11.435	$6v_1 + v_3$
11.570	$7v_1$

are only allowed for totally symmetric modes. Therefore, the Eden assignments are disputed and reassignment of modes with b_{3g} and b_{1u} symmetry in favour of a combination band assignment where both modes have a_g symmetry are recommended. Consequently, these bands have been reassigned as the v_3 mode and the combination band ($v_3 + v_2$), respectively. The average spacing between the second of the two peaks from the main progression is reported by Eden as 0.152 eV or 1225 cm^{-1} and assigned to v_{11} [10]. However, the average difference in this work between this progression and the corresponding members of the v_1 progression is 1245 cm^{-1}, but the difference between the band origin and the first member of this progression at 10.273 eV is 0.163 eV or 1315 cm^{-1}. This second value is least affected by anharmonicity and is preferred over the average value. It is also in excellent agreement with the sum of the experimental values for $v_3 + v_2$, 1315 cm^{-1}.

Thanks to the enhanced resolution, an additional, previously unobserved progression has been identified with two peaks in each member of the main progression with a spacing of 637 and 1008 cm^{-1} from the band origin. This progression with members at 10.188 and 10.235 eV is well reproduced when populating the v_6 mode (C–F$_2$ scissor out of synchronicity) of b$_{1u}$ symmetry with one and two quanta. Both the v_6 mode and the other possibility, v_9 of b$_{2u}$ symmetry, are non-totally symmetric, so should be forbidden transitions in the absence of a geometry change upon ionization. Yet by evaluating the actual nuclear wave function overlap, Franck–Condon simulations show that v_6 is populated with non-negligible intensity even without a change in symmetry. Finally, the same pattern of peaks at v_2, v_6, v_3, $2v_6$ and ($v_2 + v_3$) is repeated for each member of nv_1 ($n = 0$–6).

It appears that with the exception of 1,1-C$_2$H$_2$F$_2^+$, the rest of the series, C$_2$H$_3$F$^+$, C$_2$HF$_3^+$ and C$_2$F$_4^+$ remain planar upon ionization in the ground electronic cation state. The experimental spectra cannot be faithfully reproduced with a non-planar ion geometry in these latter three ions. Apparently, the vibronic coupling between the π(C=C) and π(C–X$_2$) where X = H or F [26], is only measurable in 1,1-C$_2$H$_2$F$_2^+$, in which a torsional twist is observed.

A final note upon the TPES of the ground electronic states of the four fluorinated ethenes presented in this work is made. The successful application of fitting the ground state band with the Franck–Condon intensities *and* apparent similarities between the PES [2, 4, 7, 8] and TPES of this work indicates that no autoionization effects are seen.

6.3.2 Electronically Excited Cation States

6.3.2.1 Spectroscopy

Equation-of-motion coupled cluster singles and doubles for ionized states, EOM-IP-CCSD/cc-pVTZ calculations [29] were undertaken using Q-Chem 3.2 [30] at the optimized G3B3 neutral geometries to determine accurate vertical ionization energies and assign excited electronic state TPES bands, the results are given in Table 6.10. Excited state wave functions of the cation are of single determinant character and Koopmans' theorem [31] holds. Thus, the EOM-IP-CCSD assignment agrees exactly with the semi-empirical HAM/3-based ordering of the cations published thirty years ago by Bieri et al. [2] and, with the exception of the almost

Table 6.10 Comparison of the calculated and experimentally observed vertical ionization energies of the excited states of $C_2H_3F^+$, 1,1-$C_2H_2F_2^+$, $C_2HF_3^+$ and $C_2F_4^+$ with literature values

Cation	State	This work			Others	
		Theory energy (eV) EOM-IP-CCSD/cc-pVTZ	Expt. TPES (eV)	Locht [7, 8] (TPES), Eden [10] (He I)	Sell and Kuppermann [4] (He I)	Bieri [2] (He II)
$C_2H_3F^+$	$\tilde{X}^2A''$	10.62	10.556 (10.364^a)	10.57	10.56	10.36^a
	$\tilde{A}^2A'$	13.84	13.76	13.76	13.80	13.8
	$\tilde{B}^2A'$	14.74	14.55	14.56	14.54	14.5
	$\tilde{C}^2A'$	16.23	16.63	16.64	16.68	16.7
	$\tilde{D}^2A''$	16.90	16.63	17.84	18.0	16.7
	$\tilde{E}^2A'$	18.12	17.83			17.9
	$\tilde{F}^2A'$	20.54	20.16	20.18		20.2
1,1-$C_2H_2F_2$	$\tilde{X}^2B_1$	10.69	10.496 (10.303^a)	10.298^a	10.69	10.29^a
	$\tilde{A}^2B_2$	15.06	14.81	14.085	14.83	14.9
	$\tilde{B}^2A_1$	15.70	15.73	15.716	15.73	15.8
	$\tilde{C}^2B_2$	15.95			15.73	16.1
	$\tilde{D}^2A_2$	16.16	(16.13^b)		15.73	16.1
	$\tilde{E}^2B_1$	18.28		18.157	18.18	18.2
	$\tilde{F}^2A_1$	18.32	18.17 19.64			18.2
	$\tilde{G}^2B_2$	19.80	19.63	19.58		
	$\tilde{H}^2A_1$	21.73	21.47	21.40		
$C_2HF_3^+$	$\tilde{X}^2A''$	10.57	10.544 (10.138^a)		10.54	10.14
	$\tilde{A}^2A'$	14.75	14.67		14.62	14.7
	$\tilde{B}^2A'$	15.93	15.9		15.90	16.0
	$\tilde{C}^2A''$	16.52	16.7		16.36	16.5
	$\tilde{D}^2A'$	16.68			16.69	16.8
	$\tilde{E}^2A''$	16.74			18.06	16.8
	$\tilde{F}^2A'$	18.07	18.0		18.56	18.0
	$\tilde{G}^2A''$	18.74	18.47			
	$\tilde{H}^2A'$	19.88	20.01			
	$\tilde{I}^2A'$	20.33	20.01			
	$\tilde{J}^2A'$	22.12	21.87			
$C_2F_4^+$	$\tilde{X}^2B_{3u}$	10.60	10.535 (10.110^a)	15.93^c	10.56	10.14
	$\tilde{A}^2B_{3g}$	15.96	15.99		15.95	15.9
	$\tilde{B}^2A_g$	16.33	16.20	16.64^c	16.63	16.6
	$\tilde{C}^2B_{2u}$	16.50	16.56	16.64^c	16.63	16.6
	$\tilde{D}^2A_u$	16.86	16.92	16.64^c	16.63	16.6

(continued)

Table 6.10 (continued)

Cation	State	This work		Others		
		Theory energy (eV) EOM-IP-CCSD/cc-pVTZ	Expt. TPES (eV)	Locht [7, 8] (TPES), Eden [10] (He I)	Sell and Kuppermann [4] (He I)	Bieri [2] (He II)
	$\widetilde{E}^2B_{1g}$	17.53	17.62	16.64[c]		16.6
	$\widetilde{F}^2B_{2u}$	18.21	18.41	17.60[c]	17.60	17.6
	$\widetilde{G}^2B_{2g}$	19.43	19.48			18.2
	$\widetilde{H}^2B_{3u}$	19.57	19.48			19.4
	$\widetilde{I}^2B_{3g}$	20.93	21.05			19.4
	$\widetilde{J}^2B_{2u}$	21.26	21.05			21.0
	$\widetilde{K}^2A_g$	22.59	22.55			21.0

[a] Adiabatic IE
[b] Shoulder
[c] Eden et al. [10]

degenerate $\widetilde{E}$, $\widetilde{F}$ and $\widetilde{H}$, $\widetilde{I}$ electronic states in 1,1-$C_2H_2F_2^+$ and $C_2F_4^+$, respectively, also with their Green's function analysis. Counter-intuitively, electronic excited state assignments are straightforward for the fluorinated ethene ions with little static electron correlation, in sharp contrast with the vibrational assignments of the ground state spectra, where it has been found that even the most recent vibrational assignments need revision [7, 8, 10].

Two of the molecules studied, C_2H_3F and C_2HF_3, have C_S symmetry, 1,1-$C_2H_2F_2$ has C_{2v} and C_2F_4 has D_{2h} symmetry. In order to establish trends and trace the evolution of the electronic ion states in the series, the Kohn–Sham orbital character symmetries according to their C_{2v} character were considered, even for the C_S molecules, as follows. Orbitals without a nodal plane along the C=C axis are classified as totally symmetric, giving the corresponding ion state A_1 symmetry. Ionization from orbitals with a nodal plane in the molecular plane but without one perpendicular to it leads to B_1 ion states. Orbitals with a nodal plane perpendicular to the molecular plane along the C=C axis correspond to B_2 ion states. When both nodal planes are present in orbitals, ionization leads to A_2 states. This assignment is shown together with the overall TPES in Fig. 6.6. The slight destabilization of the π-type HOMO corresponding to the ground ion state and the overall stabilization of the deeper lying orbitals, i.e. progressively higher ionization energies corresponding to excited ion states, with increasing F substitution confirms not only the perfluoro effect [3], but also the enhanced stabilization of the fluorine lone pairs. With the exception of this point, other trends with increasing fluorine substitution are difficult to establish.

Vibrational progressions have been observed in some excited states in all four molecules. When vibrational structure is observed in the excited states, Franck–Condon factors have to be significant for levels in the bottom of the potential energy well, and it can be assumed that the geometries of the excited state ion are

6.3 Results and Discussion

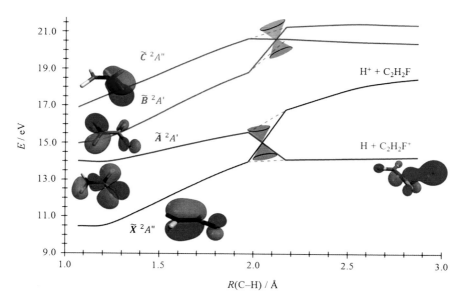

Fig. 6.6 Reaction curves of the $\widetilde{X}$, $\widetilde{A}$, $\widetilde{B}$ and $\widetilde{C}$ excited states of $C_2H_3F^+$ along the H-loss coordinate in the ground state ion. Conical intersections (indicated as funnels) are indicated at R(C–H) ≈ 2.05 Å between the $\widetilde{X}$ and $\widetilde{A}$ states, and at R(C–H) ≈ 2.1 Å between the $\widetilde{B}$ and $\widetilde{C}$ states

comparable to that of the ground neutral state. Within this approximation, only tentative assignments based on the ion ground state calculated frequencies can be made. A strong well-resolved progression is seen on the $\widetilde{H}$ $^2A'$ state of $C_2HF_3^+$ between 19.4 and 20.3 eV which has not been previously reported. The observed spacing of ca. 847 cm^{-1} could be attributed to an asymmetric wagging mode with a F_a–C_1F_b symmetric stretch, or even quanta of a F_a–C_1 = C_2 bending mode. There is very weak vibrational structure seen between 22.5 and 24.9 eV with a separation of ca. 240 cm^{-1}, which could be attributed to the CHF = CF_2 wagging mode. The strongest vibrational structure in the excited states of $C_2F_4^+$ is seen on the $\widetilde{E}$ $^2B_{1g}$ peak at 17.6 eV, also observed in the HeI spectra of Brundle et al. [3] and Eden et al. [10] Brundle et al. assign the complex structure to two separate progressions, v_2 and v_3 [3]. The v_2 mode involving the C=C stretch, makes a more likely candidate for the major progression where a vibrational spacing of ca. 777 cm^{-1} rather than the v_{11} mode (calculated at 536 cm^{-1} in the ground cation state) proposed by Eden et al. is observed [10]. The minor progression has an observed vibrational spacing of ca. 398 cm^{-1} and is assigned to the v_3 mode, a C=C stretch with symmetric C–F_2 scissor, in accordance with Brundle et al. [3]. A final single vibrational progression is seen on the peak at 19.1–19.7 eV (ionic states with $\widetilde{G}$ $^2B_{2g}$ and $\widetilde{H}$ $^2B_{3u}$ symmetry) and is assigned to the v_2 mode, ca. 777 cm^{-1}, again in accordance with Brundle et al. [3] Locht et al. [7, 8] reported a large intensity ratio for the excited vs. ground electronic state bands in the TPES, whereas in this chapter, it is found that the ground state bands have comparable intensities to the first excited state band in the TPES. This could be due to disparities between how the photon flux is accounted for.

6.3.2.2 Dynamics

The dissociative photoionization dynamics is of both applied and fundamental interest [32]. On the one hand, appearance energies can be used in thermochemical derivations, but only if the dissociative photoionization is fast at the thermochemical threshold [33] or if the dissociation rates can be measured and extrapolated to it [34]. In addition to new and accurate neutral thermochemistry, such thresholds can also help interpret the products of ion–molecule bimolecular reactions [35]. On the other hand, understanding the energy flow between different electronic and nuclear degrees of freedom is of paramount fundamental interest. A dissociation process in any molecular system (neutral or charged) is typically considered statistical if the intermediate state is sufficiently long-lived to allow for the complete redistribution of internal energy before dissociation. Such processes are dominated by the ground electronic state, since its density of states exceed that of any excited state by orders of magnitude [32]. Non-statistical, non-ergodic processes are characterized by an incomplete sampling of the energetically allowed phase space of the dissociating species. The reason can be a fast dissociation process, such as impulsive F-loss from CF_4^+ [36], or Cl-loss from CCl_4^+ [37]. Alternatively, an electronically excited ion state can be so long-lived so that it establishes a second dissociation regime, shielded from access to the ground state dynamics of that surface. This was shown to be the case in F-atom loss in $C_2F_4^+$ [11], and probably applies in CH_3-loss from CH_3OH^+ [38]. The nature of the non-statistical fluorine atom loss from singly to triply fluorinated ethene cations has long been misunderstood [12–15]. Contrary to F-loss from $C_2F_4^+$, detailed kinetic energy release studies have shown that F-loss from $C_2H_3F^+$ and 1,1-$C_2H_2F_2^+$ is, in part, an impulsive process [13, 39]. The threshold F-loss ion yield curves were shown to correlate only approximately with the TPES signal [11], which indicates a complex mechanism with possible Rydberg-state involvement. In contrast to $C_2F_4^+$, the state which leads unhindered to F-loss is not the first electronically excited state of the parent ion in the other members of the series. Intermediate Rydberg and ion states facilitate fast internal conversion and rule out long-lived electronically excited states. Therefore, non-statistical F-loss channels have to be fast and impulsive.

In the context of the overall valence TPES and experimental and computational information on excited electronic state energetics presented herein, it is now possible to further the discussion on the unimolecular dissociation dynamics of fluorinated ethylenes with respect to a previous study (see Chap. 5) [11]. Out of computational practicality, only the dissociation mechanism of monofluoroethene cations is considered, using EOM-IP-CCSD/cc-pVTZ calculations along the optimized [11] cation ground electronic state H-, HF- and F-loss reaction paths.

That is, the reaction path geometries are optimized at the B3LYP/6-311 ++G(d,p) level on the ground cation state, and vertical excitations are considered to the electronically excited states. Because of the spectral similarity, an analogous mechanism is expected to apply to the dissociation dynamics of di- and trifluoroethene cations. By contrast, the spectral sparsity of the TPES of tetrafluoroethene leads to a de-coupling of the $\tilde{A}$ ion state from the $\tilde{X}$ ion state, leading to isolated-state behaviour with long-lived $\tilde{A}$ state intermediates [11]. By understanding the role of different electronically excited ion states in the mechanism of the main dissociative photoionization channels, it will be shown how and why F-atom loss assumes a non-statistical character in higher internal energy states of the parent ions, whereas the HF- and H-loss channels do not.

6.3.2.3 H-atom Loss from $C_2H_3F^+$

The H-loss reaction energy curves in the monofluoroethene cation are shown in Fig. 6.2. The doublet ground ion state is of A'' symmetry. The molecular orbital of the missing electron in the dominant electron configuration of the cation in the first four electronic states is also shown in the figure, together with that in the electronic ground state of the ion, at a C–F bond length of 2.9 Å. The F atom is pointing out of the plane towards the reader in the schematic structures. The ground state of the H-loss fragment ion, $CH_2 = CF^+$, is closed shell, i.e. totally symmetric (A') in C_S symmetry and the spin density is localized in the 1 s orbital of the leaving H atom in the products.

In other words, the $\tilde{X}$ state of the parent ion adiabatically correlates with the $CH_2 = CF + H^+$ dissociation products, but H^+ is not observed in the valence photoionization experiments [11]. The thermochemical threshold to H-atom loss is 13.6 eV [11]. In order to determine the well depth of the $\tilde{A}$ state, which can indeed correlate adiabatically with the H-loss products, an EOM-IP-CCSD geometry optimization was carried out, that yielded a structure with an elongated α-C–H bond length and increased C–C–F bond angle as well as an adiabatic ionization energy of 13.18 eV. Therefore, the $\tilde{A}$ state of $C_2H_3F^+$ is bound by ≈400 meV, it adiabatically correlates with the ground state H-loss products, and is coupled with the $\tilde{X}$ state through the C=C–F bend coordinate. Fast relaxation through this conical intersection leads to statistical H-loss with $k > 10^7$ s^{-1} at threshold. Contrary to the diabatic coupling coordinates in HF and F losses (see later), the coupling vibrational mode in this case is the H–C=C bending mode, and not the reaction coordinate. This explains the discontinuities in the potential energy curves plotted in Fig. 6.2. In the ground electronic state constrained geometry optimizations, electronic state switching occurs at a reaction coordinate value, at which the new state is more stable even at the H–C–C bond angle of the original state. Thus, there is a discontinuity in this bond angle and the curve crossings do not correspond to a point along the seam of the conical intersection. Instead, the seam is only known to be located within the dashed lines. As also seen in Fig. 6.2, the $\tilde{B}$

and $\widetilde{C}$ states are also coupled by a conical intersection but are distinct from the $\widetilde{X}$ and $\widetilde{A}$ states. Thus, only the $\widetilde{X}$ and $\widetilde{A}$ states participate in the H-loss channel with the exit channel being the $\widetilde{A}$ state.

6.3.2.4 HF Loss from $C_2H_3F^+$

The HF-loss potential energy curves are shown in Fig. 6.7. After the neutral and closed shell HF leaves, the spin density is localized in the π-system of the ethyne fragment ion. The reaction coordinate is taken as the distance between the midpoints of the C=C and H–F bonds, and, again, there appears to be a conical intersection at play at $R \approx 1.5$ Å. The potential energy curves cross smoothly, because the coupling vibrational coordinate is very similar to the reaction coordinate of choice. However, the $\widetilde{X}\ ^2A''$ and $\widetilde{A}\ ^2A'$ curves are degenerate at the dissociation limit ($R \gg 2.5$ Å, not shown in Fig. 6.7), as they only differ in the orientation of the degenerate ethyne π-orbitals from which the electron is removed from to form $C_2H_2^+$.

As one approaches the products, the (HOMO–1) of the neutral takes a σ-antibonding character between the fragments, whereas the HOMO corresponds to a π-antibonding orbital. Thus at $R = 2$ Å, the A' state of the cation with one σ-antibonding electron removed is more stable than the A'' state ion (see schematic

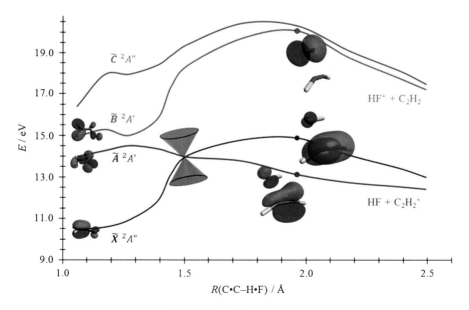

Fig. 6.7 Reaction curves of the $\widetilde{X}$, $\widetilde{A}$, $\widetilde{B}$ and $\widetilde{C}$ excited states of $C_2H_3F^+$ along the HF-loss coordinate. A conical intersection has been located at R(C·C–H·F) ≈ 1.5 Å

structures in the figure with the corresponding molecular orbitals in the neutral at the highlighted reaction coordinate with the HF leaving upwards). The energy difference between the two states is larger than 0.5 eV when HF is removed by 2.5 Å from $C_2H_2^+$, showing a long range interaction at play. The $\widetilde{B}\ ^2A'$ state of $C_2H_3F^+$ correlates with the ground state $C_2H_2 + HF^+$ products, lying 5 eV above the lower energy $\widetilde{X}$ and $\widetilde{A}$ channels, the corresponding molecular orbital of the missing electron being of F lone pair character, as shown. On the $\widetilde{B}$ state surface, the transition state to HF loss lies 20 eV above the neutral, and is even higher for the $\widetilde{C}$ and $\widetilde{D}$ ($\widetilde{D}$ state not shown) states which correlate with excited state products. As was the case for H loss, HF loss in $C_2H_3F^+$ is related by the interplay between the $\widetilde{X}$ and $\widetilde{A}$ states. Higher-lying ion states are de-coupled from the HF-loss channel observed in dissociative valence photoionization, as they must first relax to the $\widetilde{X}/\widetilde{A}$ manifold in order to lose HF in a statistical fashion.

6.3.2.5 F-Atom Loss from $C_2H_3F^+$

The F-loss potential energy curves, shown in Fig. 6.8, show a different pattern. The first three ion states, $\widetilde{X}$, $\widetilde{A}$ and $\widetilde{B}$, dissociate to products with different singly occupied fluorine $2p_{x,y,z}$ orbitals and the same ground electronic state of the

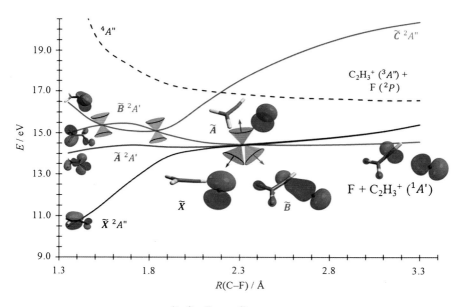

Fig. 6.8 Reaction curves of the $\widetilde{X}$, $\widetilde{A}$, $\widetilde{B}$ and $\widetilde{C}$ excited states of $C_2H_3F^+$ along the F-loss coordinate. Three conical intersections have been found R(C–F) ≈ 1.5 Å between $\widetilde{C}$ and $\widetilde{B}$, R(C–F) ≈ 1.8 Å between $\widetilde{B}$ and $\widetilde{C}$ and at R(C–F) ≈ 2.3 Å between the $\widetilde{X}$, $\widetilde{A}$, and $\widetilde{B}$ states leading to degenerate asymptotes

CH_2CH^+ ion. At a fluorine–carbon distance of around $R(C–F) = 2.3$ Å, the three states are degenerate and are coupled by the C–F stretch coordinate. At longer distances, the $\tilde{B}$ state appears to be converged to the dissociation limit, whereas even at $R = 3.3$ Å, the $\tilde{X}$ and $\tilde{A}$ states increase in energy. This suggests that long range interactions are significant at even longer distances than for HF-loss, as there has to be an $\tilde{X}/\tilde{A}$ F-loss transition state at $R(C–F) > 3.3$ Å. To describe this bond length region reliably, the triple-ζ basis set used in these calculations would need to be augmented with several diffuse functions.

The non-statistical F-loss process arises in the energy range of the $\tilde{C}^2A''$ state. Based on the low energy component in the kinetic energy release distribution and ab initio calculations, the dissociation channel from this $\tilde{C}$ state to the $C_2H_3^+$ ($^3A''$) + F (^2P) products was suggested to play an important role [7, 14, 15]. Rooda et al. [15] further established that the $\tilde{C}$ state has a large negative energy gradient towards the C–F elongation, and suggested a diabatic pathway in which the $\tilde{C}$ and $\tilde{X}$ states couple through an avoided crossing at $R(C–F) = 2.0$ Å with a minimum energy gap of around 0.96 eV. Their proposed route to F-atom formation is either via this diabatic pathway along which the initial momentum in the C–F stretch is retained, leading to ground state $C_2H_3^+$ ($^1A'$) with a large translational kinetic energy release, or via an adiabatic pathway along the $\tilde{C}$ state producing electronically excited triplet $C_2H_3^+$ ($^3A''$) fragment ion with small kinetic energy release. The fragments of the latter channel, however, correspond to a quartet wave function, meaning that it cannot be the asymptote of the doublet $\tilde{C}$ state. Indeed, it was found that the $^4A''$ quartet state crosses the $\tilde{C}^2A''$ state at $R(C–F) \approx 2.2$ Å (Fig. 6.8). The rate of the intersystem crossing is, however, unlikely to exceed that of internal conversion to lower lying doublet ion states. The breakdown diagram [11] casts further doubt on the feasibility of this pathway. The CBS-APNO [40] calculated splitting between the singlet and triplet states of the vinyl cation is 2.10 eV, putting the asymptote to triplet $C_2H_3^+$ production at ≈ 16.1 eV (cf. the 16.6 eV limit in Fig. 6.8 corresponding to the triplet energy at the singlet $C_2H_3^+$ product geometry, calculated using EOM-IP-CCSD/cc-pVTZ). In the breakdown diagram of $C_2H_3F^+$ [11], the percentage yield of F-loss production plateaus at ≈ 30 % in the statistical regime, then rises rapidly in the photon energy region 15.5–16 eV to a constant level of ≈ 60 % from which the signal decreases above a photon energy of 17.0 eV. If triplet $C_2H_3^+$ production were a viable dissociation path, the F-loss signal should increase above its threshold at 16.1 eV, at which energy it has already reached its asymptotic value. The absence of such an increase rules out significant $C_2H_3^+$ ($^3A''$) production and an alternative mechanism is offered below.

As opposed to the $\tilde{X}$, $\tilde{A}$ and $\tilde{B}$ states, the $\tilde{C}$ state of $C_2H_3F^+$ converges to $CH_2CH + F^+$, and does not lead to F-loss products. However, it is just below the onset of this excited A'' ion state peak in the threshold photoelectron spectrum [11] that the non-statistical and partly impulsive [13] F-loss channel opens up.

6.3 Results and Discussion

Figure 6.8 shows that the $\tilde{B}$ and $\tilde{C}$ states are coupled at 1.5 Å $< R$(C–F) $<$ 1.8 Å. As Rooda et al. have shown [15], the C–F bond length in the $\tilde{C}$ state minimum is markedly longer than in the ground state of the ion, thanks to the removal of an electron from a π-type C–F bonding orbital in ionization to the $\tilde{C}$ ion state. Consequently, $\tilde{C}$ state ions are highly vibrationally excited in the Franck–Condon envelope with large excitation in modes associated with the C–F stretch. $\tilde{C}$ state ions can lose electronic excitation energy by crossing through the conical intersections to the $\tilde{B}$ state. If the crossing occurs at low C–F bond lengths on the bound part of the $\tilde{B}$ surface, the resulting species will decay statistically. However, at higher C–F bond lengths, the $\tilde{B}$ state also has a repulsive character that facilitates F-atom loss. The fate of the parent cation is still not sealed at this point, since fluorine p-orbital mixing at R(C–F) = 2.3 Å can lead it onto the partially bound $\tilde{A}$ and $\tilde{X}$ states, yielding a longer lived F-loss intermediate, in which redistribution of the excess energy may, to a certain extent, still be possible. Thus, three different F-loss channels are proposed in the $\tilde{C}$ state band of the TPES of C_2H_3F: (i) statistical F-loss mostly from the $\tilde{X}$ state by a $\tilde{C} \rightarrow \tilde{B}$ transition on the bound part of the $\tilde{B}$ state surface, through the first conical intersection, (ii) impulsive F-loss by a $\tilde{C} \rightarrow \tilde{B}$ transition onto the repulsive part of the $\tilde{B}$ state and subsequent direct dissociation, and, as a slight variation of this process, (iii) semi-impulsive non-statistical F-loss by a multi-step $\tilde{C} \rightarrow \tilde{B} \rightarrow \tilde{X}/\tilde{A}$ transition with an intermediate at R(C–F) $\approx$ 2.3 Å.

This multiple channel F-loss mechanism explains the proposed bimodal kinetic energy release distribution observed for $C_2H_3F^+$ and 1,1-$C_2F_2H_2^+$ [14, 15, 39], as the low kinetic energy release modus is a result of the statistical dissociation pathway. Furthermore, direct $\tilde{C}$-state involvement is not necessarily required in threshold photoionization. As was proposed in the study of CH_3I [23], the neutral parent can be excited to the Rydberg manifold in the initial step. The Rydberg manifolds belonging to each ion state will have the similar characteristics to the ion state, and autoionization may also occur after internal conversion. This explains why the non-statistical F-loss channel is seen at slightly lower energies than the actual $\tilde{C}$-state peak in the photoelectron spectrum of $C_2H_3F^+$. In the iodomethane study [23], neutral dissociative states were proposed to connect different Rydberg manifolds with the corresponding ion states lying approximately 2 eV apart. In monofluoroethene, such neutral states do not need to be invoked, since the Rydberg manifolds themselves can readily interconvert at conical intersections. Such conical intersections may play a significant and, as yet, unrecognized role in ensuring that most molecules with a sufficiently congested ion spectrum dissociatively photoionize in accordance with statistical theory [41].

6.4 Conclusions

The ground state TPES of four fluorinated ethenes; C_2H_3F, $1,1\text{-}C_2H_2F_2$, C_2HF_3 and C_2F_4 have been recorded at a higher resolution than previously reported. The ground state spectra have been simulated and fitted using the Franck–Condon fitting program, FCfit, to better identify those vibrational modes active upon ionization [1]. A number of weak peaks seen in the ground state band of $C_2H_3F^+$ have been reassigned. Even quanta transitions of the v_{12} mode are allowed and $2v_{12}$ contributions have been identified. Also, the vibrational transitions in the ground electronic state TPES of $1,1\text{-}C_2H_2F_2$ were reassigned. In addition to the v_2 C=C stretching mode previously observed by others, the v_3 mode was identified which gives rise to asymmetry of all of the peaks in the v_2 progression. The Franck–Condon analysis has also yielded a surprising result, revealing a small geometry change upon ionization. The loss of planarity arises through the vibronic coupling of the ground and first excited electronic states via the torsional vibrational mode, suggesting a conical intersection is at play. This led to the assignment of a non-symmetric v_{10} (a_2) mode apparent in odd quanta. By contrast, Franck–Condon analysis shows that planar geometries in the monofluoroethene, trifluoroethene and tetrafluoroethene ions are retained. The ground state TPES of C_2HF_3 has been recorded with significantly improved resolution than in previous studies. The v_9, v_8, v_7, v_5 and v_4 vibrational progressions have been identified in addition to the v_2 C=C stretching mode previously identified by Sell and Kupperman [4]. Finally, the vibrational progressions in the C_2F_4 ground state TPES have been extensively re-assigned from the HeI study of Eden et al. [10]. In addition to the strong C=C stretching mode nv_1 observed previously, weak progressions are assigned to the v_3, v_2 and $(v_3 + v_2)$ vibrational modes, all with a_g symmetry.

Excited state threshold photoelectron spectra are also reported for the four fluoroethenes up to 23 eV together with the computed vertical ionization energies. In contrast to the ground-state vibrational assignments, historical electronic state assignments have been found to be remarkably accurate. Based on excited state calculations on $C_2H_3F^+$ and new experimental data included in Chap. 5, a new model is proposed for the non-statistical dissociative photoionization decay mechanism by F-atom loss as well as the previously observed bimodal F-loss kinetic energy release distribution. Triplet $C_2H_3^+$ fragment ion production by intersystem crossing is ruled out in the new mechanism, as is the isolated state mechanism proposed for F-loss from $C_2F_4^+$, in which the large separation of the electronic states slows down internal conversion. Instead, the $\widetilde{C}^2A''$ state of $C_2H_3F^+$ acts as a gateway with conical intersections to bound and dissociative parts of the $\widetilde{B}$ state potential energy surface. Together with H and HF loss, statistical F-loss takes place via the $\widetilde{X}$ state, whereas diabatic coupling onto the repulsive part of the $\widetilde{B}$ state surface is responsible for non-statistical, impulsive F-loss.

References

1. Spangenberg, D., Imhof, P., & Kleinermanns, K. (2003). *Physical Chemistry Chemical Physics: PCCP, 5,* 2505–2514.
2. Bieri, G., Niessen, W. V., Åsbrink, L., & Svensson, A. (1981). *Chemical Physics, 60,* 61–79.
3. Brundle, C. R., Robin, M. B., Kuebler, N. A., & Harold, B. J. (1972). *Journal of the American Chemical Society, 94,* 1451–1465.
4. Sell, J. A., & Kuppermann, A. J. (1979). *Chemical Physics, 71,* 4703–4715.
5. Takeshita, K. (1999). *Theoretical Chemistry Accounts, 101,* 343–351.
6. Takeshita, K. (1999). *Chemical Physics, 250,* 113–122.
7. Locht, R., Leyh, B., Dehareng, D., Hottmann, K., & Baumgärtel, H. (2010). *Journal of Physics B: Atomic, Molecular and Optical, 43,* 015102–015117.
8. Locht, R., Dehareng, D., & Leyh, B. (2012). *Journal of Physics B: Atomic, Molecular and Optical, 45,* 115101–115118.
9. Jarvis, G. K., Boyle, K. J., Mayhew, C. A., & Tuckett, R. P. (1998). *Journal of Physical Chemistry A, 102,* 3230–3237.
10. Eden, S., Limão-Vieira, P., Kendall, P. A., Mason, N. J., Delwiche, J., Hubin-Franskin, M.-J., et al. (2004). *Chemical Physics, 297,* 257–269.
11. Harvey, J., Bodi, A., Tuckett, R. P., & Sztáray, B. (2012). *Physical Chemistry Chemical Physics: PCCP, 14,* 3935–3948.
12. Dannacher, J., Schmelzer, A., Stadelmann, J.-P., & Vogt, J. (1979). *International Journal of Mass Spectrometry and Ion Physics, 31,* 175–186.
13. Güthe, F., Locht, R., Leyh, B., Baumgärtel, H., & Weitzel, K.-M. (1999). *Journal of Physical Chemistry A, 103,* 8404–8412.
14. Momigny, J., & Locht, R. (1993). *Chemical Physics Letters, 211,* 161–165.
15. Roorda, M., Lorquet, A. J., & Lorquet, J. C. (1991). *Journal of Physical Chemistry, 95,* 9118–9121.
16. Chang J.-L., Huang C.-H., Chen S.-C., Yin T.-H., & Chen, Y.-T. (2012). *Journal of Computer Chemistry.* doi: 10.1002/jcc.23194.
17. Hemberger, P., Noller, B., Steinbauer, M., Fischer, I., Alcaraz, C., Cunha de Miranda, Br K, et al. (2010). *Journal of Physical Chemistry A, 114,* 11269–11276.
18. Koziol, L., Mozhayskiy, V. A., Braams, B. J., Bowman, J. M., & Krylov, A. I. (2009). *Journal of Physical Chemistry A, 113,* 7802–7809.
19. Willitsch, S., Hollenstein, U., & Merkt, F. (2004). *Journal of Chemical Physics, 120,* 1761–1774.
20. Köppel, H., Cederbaum, L. S., & Domcke, W. (1982). *Journal of Chemical Physics, 77,* 2014–2022.
21. Sannen, C., Raşeev, G., Galloy, C., Fauville, G., & Lorquet, J. C. (1981). *Journal of Chemical Physics, 74,* 2402–2412.
22. Chupka, W. A. (1993). *Journal of Chemical Physics, 98,* 4520–4531.
23. Bodi, A., Shuman, N. S., & Baer, T. (2009). *Physical Chemistry Chemical Physics: PCCP, 11,* 11013.
24. Morse, P. M. (1929). *Physical Review, 34,* 57–64.
25. Herzberg, G., & Teller, E. (1933). *Zeitschrift für Physikalische Chemie, Abteilung B, 21,* 410.
26. Köppel, H., Cederbaurn, L. S., Domcke, W., & Shaik, S. S. (1983). *Angewandte Chemie (International ed. in English), 22,* 210–224.
27. Shimanouchi, T. (1972). Tables of Molecular Vibrational Frequencies, Consolidated (Vol. I. pp. 1–160). National Bureau of Standards.
28. Mulliken, R. S. (1955). *Journal of Chemical Physics, 23,* 1997–2011.
29. Landau, A., Khistyaev, K., Dolgikh, S., & Krylov, A. I. (2010). *Journal of Chemical Physics, 132,* 014109–001422.

30. Shao, Y., Molnar, L. F., Jung, Y., Kussmann, J., Ochsenfeld, C., Brown, S. T., Gilbert, A. T. B., Slipchenko, L. V., Levchenko, S. V., O'Neill, D. P. Jr, R. A. D., Lochan, R. C., Wang, T., Beran, G. J. O., Besley, N. A., Herbert, J. M., Lin, C. Y., Voorhis, T. V., Chien, S. H., Sodt, A., Steele, R. P., Rassolov, V. A., Maslen, P. E., Korambath, P. P., Adamson, R. D., Austin, B., Baker, J., Byrd, E. F. C., Dachsel, H., Doerksen, R. J., Dreuw, A., Dunietz, B. D., Dutoi, A. D., Furlani, T. R., Gwaltney, S. R., Heyden, A., Hirata, S., Hsu, C.-P., Kedziora, G., Khalliulin, R. Z., Klunzinger, P., Lee, A. M., Lee, M. S., Liang, W., Lotan, I., Nair, N., Peters, B., Proynov, E. I., Pieniazek, P. A., Rhee, Y. M., Ritchie, J., Rosta, E., Sherrill, C. D., Simmonett, A. C., Subotnik, J. E., III, H. L. W., Zhang, W., Bell, A. T., Chakraborty, A. K., Chipman, D. M., Keil, F. J., Warshel, A., Hehre, W. J., III, H. F. S., Kong, J., Krylov, A. I., Gill, P. M. W., & Head-Gordon, M. (2006). *Physical Chemistry Chemical Physics: PCCP, 8*, 3172–3191.
31. Koopmans, T. (1934). *Physica, 1*, 104–113.
32. Baer, T., Guerrero, A., Davalos, J. Z., & Bodi, A. (2011). *Physical Chemistry Chemical Physics: PCCP, 13*, 17791.
33. Harvey, J., Tuckett, R. P., & Bodi, A. (2012). *Journal of Physical Chemistry A, 116*, 9696–9705.
34. Bodi, A., Brannock, M. D., Sztáray, B., & Baer, T. (2012). *Physical Chemistry Chemical Physics: PCCP, 14*, 16047–16054.
35. Simpson, M. J., & Tuckett, R. P. (2012). *Journal of Physical Chemistry A, 116*, 8119–8129.
36. Simm, I. G., Danby, C. J., Eland, J. H. D., & Mansell, P. I. (1976). *Journal of the Chemical Society, Faraday Transactions 2, 72*, 426–434.
37. Smith, D. M., Tuckett, R. P., Yoxall, K. R., Codling, K., & Hatherly, P. A. (1993). *Chemical Physics Letters, 216*, 493–502.
38. Borkar, S., Sztáray, B., & Bodi, A. (2011). *Physical Chemistry Chemical Physics: PCCP, 13*, 13009–13020.
39. Gridelet, E., Dehareng, D., Locht, R., Lorquet, A. J., Lorquet, J. C., & Leyh, B. (2005). *Journal of Physics Chemistry A, 109*, 8225.
40. Ochterski, J. W., Petersson, G. A., & Montgomery, J. A. (1996). *Journal of Chemical Physics, 104*, 2598–2619.
41. Baer, T., Sztáray, B., Kercher, J. P., Lago, A. F., Bodi, A., Skull, C., et al. (2005). *Physical Chemistry Chemical Physics: PCCP, 7*, 1507–1513.

Chapter 7
Conclusions and Further Work

7.1 Conclusions

The different aspects of the imaging photoelectron photoion coincidence apparatus have been utilized to investigate the fast and slow dissociation dynamics of small halogenated cations, and to explore their potential energy surfaces in the ground and excited electronic states. Throughout the work, quantum mechanical calculations have been used to model, support and enhance the experimental results and to yield new conclusions.

A holistic approach consisting of experiment, modelling and calculation was taken in the study of the dissociation dynamics of the halogenated methanes to establish updated thermochemical values at 0 K for the neutrals; CCl_4, $CBrClF_2$, $CClF_3$, CCl_2F_2 and CCl_3F, and the ions; CH_2F^+, CHF_2^+, $CClF_2^+$, CCl_2F^+, $CHCl_2^+$ and CCl_3^+ (Chap. 4). The two experimental and calculation approaches to determining thermochemical values have been combined and complement each other. This shows that, while in many instances calculations may yield thermochemical values better then experiment, for the near future, experiments are not wholly redundant, not even for small molecules such as halogenated methanes. This will be the case as long as calculations are restricted by the size and number of electrons within the system, and so by computing cost.

The fast and slow dissociation dynamics of four fluorinated ethenes have been studied (Chap. 5). The study reveals that they can be divided into two groups, the 'time bombs' (monofluoroethene and 1,1-difluoroethene) and the 'fast dissociators' (trifluoroethene and tetrafluoroethene). The first dissociation channel of the time bombs is blocked by a tight transition state, resulting in long parent ion lifetimes. When the dissociation eventually occurs, it imparts large amount of kinetic energy into the fragments. The appearance energies were modelled using the statistical rigid-activated-complex RRKM unimolecular rate theory. On the other hand, for the fast dissociators, the ion is found to undergo rearrangement before dissociation. However the rearrangement step is not rate determining, and there is no overall reverse barrier to dissociation. The group of molecules has also provided an excellent opportunity to understand how the excess energy is redistributed into the fragments. The majority of fragmentation reactions are found to be statistical processes, whereas F-atom loss is predominantly non-statistical at high internal energies. This dominant channel was explored in the context of $C_2F_4^+$, where fortuitously the non-statistical product distributions could be untangled from the rest of the statistical dissociation channels. The ion yield curves of $C_2F_3^+$ via F-loss and the subsequent sequential dissociation were modelled, using statistical theories. Together with ab initio calculations of the potential energy surfaces, it was found that the electronic state of $C_2F_4^+$ giving rise to F-atom loss was isolated, but the dissociation from it was a statistical process. That is, when only the ground state was inaccessible and the ion long-lived, then excess energy is statistically redistributed among the nuclear degrees of freedom.

The ground state TPES of the same four fluorinated ethenes was recorded at high resolution and Franck–Condon factors were calculated and fitted to the ground state photoelectron band, to yield the ion geometries (Chap. 6). The excited electronic state potential energy surfaces of $C_2H_3F^+$ were calculated to reveal the reason why the F-atom loss channel exhibited both a statistical (at lower energies) and non-statistical (at higher energies) decay mechanism, unlike the single mode of F-loss in $C_2F_4^+$. In $C_2H_3F^+$, it was found that an abundance of conical intersections was responsible for a bi-modal decay mechanism. Non Born–Oppenheimer behaviour is also suggested to be involved in the distortion of the dihedral angle of 1,1-difluoroethene upon ionization. Vibronic coupling between the ground and first excited electronic state occurs which is mediated by the torsional vibrational mode.

7.2 Further Work

The high resolution threshold photoelectron spectra of the chlorinated methanes, CH_3Cl, CH_2Cl_2 and $CHCl_3$, and the fluorinated methanes CH_3F, and CH_2F_2 were recorded during the period of study. Due to time constraints these spectra could not be fully analysed, and are presented in Appendix C.

Several questions have been raised throughout this thesis that requires further investigation. The first would be to investigate the potential energy well depths of the halogenated methane parent (and subsequent daughter) ions with coincidence and computational techniques. This study would include methanes comprised of all combinations of mixed halogen-hydrogen atoms with the addition of iodine containing species. The aim would be to answer the following question; at what point does the well become too 'shallow' to accommodate the transposition of the neutral thermal energy distribution onto the ionic manifold? Reassuringly, the general trend of stability for species containing only one type of halogen atom follows the periodic table sequence for the group of halogens. However, it is the effect that the particular combinations of mixed halogens and hydrogens within the same molecule have on ion stability which is less readily understood. This study could be expanded to investigate which other systems have shallow parent ion potential wells.

A second interesting point raised during the course of this PhD is the apparent twist in 1,1-diflouroethene upon ionization. The presence of a conical intersection, which enables the coupling between the ground and first excited electronic states of the ion and gives rise to the twisted structure, would need to be confirmed with ab initio calculations. Further calculations are required to uncover why the states interact less with increasing fluorine substitution. Modelling the Franck–Condon factors for high resolution threshold electron spectra of the other conformational isomers (1,2-*trans* and 1,2-*cis*) of difluoroethene would also be desirable, to see if they share a common potential energy hypersurface or twisted parent ion geometry. It would also be interesting to study the dissociation dynamics of these other conformational isomers, to see if their parent ions also undergo reorganization and then dissociate to the same daughter ion, thus sharing the reaction coordinate as is the case for the three isomers of $C_2H_2Cl_2^+$ [1].

7.3 Beyond TPEPICO

A natural progression from the imaging threshold photoelectron photoion (iPEP-ICO) setup is simultaneously velocity map imaging both the electrons *and* the ions giving an i^2PEPICO experiment [2, 3]. This would enable the detection of the original three-dimensional Newton sphere of the ions, which contains information about the kinetic energy and orientation of the product ions. As with TPEPICO, we would also know the internal energy of the ions because they are measured in coincidence with zero kinetic energy electrons. Furthermore, mass selection would also be possible by recording the ion time-of-flight distributions. This offers an advantage over traditional velocity map imaging, which maps the square root of the kinetic energy of all particles, $E^{1/2}$, *irrespective* of their mass to charge ratio [4], where all particles of differing mass but with the same kinetic energy are imaged on the same spot of the detector. The kinetic energy release [3], the angular distributions [5] and anisotropy parameters [3, 6] (β) of the ions as a function of photon energy could be studied.

The one-photon mode of threshold ionization, using the easily tuneable synchrotron radiation source, offers another advantage over laser-based studies which use multi-photon modes of ion preparation (e.g. resonance enhanced multi-photon ionization, REMPI). By virtue of using non-resonant excitation in tuneable synchrotron studies, indirect ionization mechanisms provide access to energy levels within Franck–Condon gaps, thereby delivering measurable ion and electron signals [7].

Preliminary studies which form the foundations of such an i^2PEPICO experiment have been performed on CF_4, CCl_4, SF_6 and SF_5CF_3. Only the results of SF_5CF_3 will be discussed here. Ion images and preliminary data analysis of the remaining three molecules can be found in Appendix D. Ion images were obtained by reversing the polarities of the detectors so that ions were drawn out towards the imaging Roentdek DLD40 position-sensitive delay-line-detector. However, due to the constraint of having only one imaging detector, it was not possible to select only those electrons with zero kinetic energy and their corresponding ions Therefore, ions associated with electrons over a range of energies were detected. Furthermore, no mass selection was possible as the ion time-of-flights could not be recorded. The experiment is therefore to be considered solely as an unrefined, 'one-sided' ion imaging experiment, where resolution is further limited as a result of using a thermal and not a molecular beam source. As such, only these species which dissociatively photo ionize into a single fragment ion over a range of eV were chosen for study. The ground electronic states of the four cation are repulsive in the Franck–Condon region, so only one fragment ion is detected at low energies. Consequently for such molecules, measuring the dissociative ionization energy (*DIE*) with conventional TPES is troublesome. By measuring the mean total kinetic energy release (TKER) of the fragments as function of photon energy, the *DIE* can be obtained by extrapolation to a TKER of zero [8].

The *DIE* can then be used to determine the adiabatic ionization energy (*AIE*) of fragment radical (A) by the relationship,

$$AB + h\nu \rightarrow A^+ + B + e^- \tag{7a}$$

$$AIE(A) = DIE(AB) - D_0(A-B) \tag{7.1}$$

where D_0 is the bond dissociation energy.

This extrapolation has been done in a previous study for CF_4, SF_6 and SF_5CF_3 by Chim et al. [8]. However the uncertainty of the adiabatic ionization energies obtained for CF_3 and SF_5 are greater than with conventional TPES. The *AIE* of the CF_3 radical obtained from that study on SF_5CF_3 of 8.84 ± 0.20 eV, is toward the lower end of the range of reported values and lower than that determined in this PhD of 9.090 ± 0.015 eV (see Chap. 5). In the method employed by Chim et al. the TKER is obtained from the TOF distributions [8]. However the energy releases are calculated from the ion TOF or the projection of the recoil velocity. So if the molecular cation dissociates anisotropically, i.e. the timescale for dissociation is significantly less than the timescale for one rotational period of the cation [9], then the TKER can be under or over estimated. Therefore an error is introduced in determining the intercept. Additional sources of error are introduced if the extrapolation is not linear and also from decreased ion TOF resolution.

By directly recording the ion hit positions on the imaging detector, the TKER, product alignment and orientation (i.e. the distribution of the product angular momentum relative to the parent transition moment and so too the electric vector of the incoming radiation) can be determined across a range of photon energies. The first step is the analysis of the two-dimensional projections of the three-dimensional fragment ion distributions (the Newton sphere of expanding charged particles) using the inversion method pBasex, developed by Garcia et al. [6]. to yield ion product flux contour maps. This inversion method of analysis was chosen instead of the onion peeling algorithm [10] because it generates less noise down the centre of the image. This is important to note, because the resolution of the

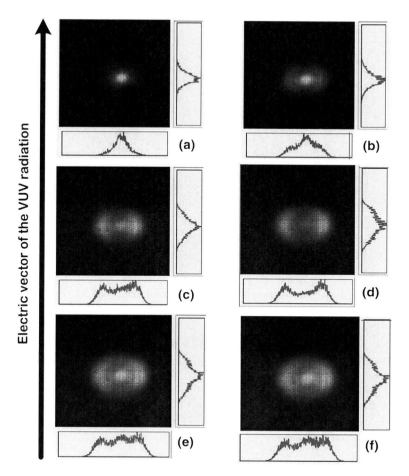

Fig. 7.1 Raw ion images of $SF_5CF_3^+$ as captured by the Roentdek DLD40 detector at the following energies **a** 13.0, **b** 13.4, **c** 13.7, **d** 14.5, **e** 14.9 and **f** 15.0 eV. The side graphs show the cross sections of the distributions through the mid-point

images is poor in comparison to those generated with a combined molecular beam and laser photon source. Therefore there is a need to limit the amount of noise generated by analysis, which is caused by an over subtraction of the contribution made by faster particles, and statistical fluctuations present towards the outer perimeter of the imagers which can be amplified [6].

SF_5CF_3 dissociates into CF_3^+ and SF_6, so only the CF_3^+ is detected,

$$SF_5CF_3 + h\nu \rightarrow CF_3^+ + SF_5 + e^- \qquad (7b)$$

The raw ion images of CF_3^+ from SF_5CF_3 are shown in Fig. 7.1. With assistance from Dr. Mick Staniforth (formerly of the University of Nottingham, now University of Warwick) and Prof. Ivan Powis (University of Nottingham), the

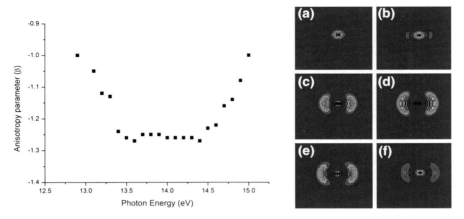

Fig. 7.2 Anisotropy parameters obtained from the reconstructed ion images across the energy range 12.9–15.0 eV. Selected reconstructions are shown for the energies; **a** 13.0, **b** 13.4, **c** 13.7, **d** 14.5, **e** 14.9 and **f** 15.0 eV, anisotropy in the images can clearly be seen. Contamination from background oxygen and nitrogen ions with a thermal energy distribution contributes to the signal at the centre of the images, whereas the lobes in images (**b**) to (**f**) are as a result of the kinetic energy release in the CF_3^+ fragments

reconstructed distributions were generated using pBasex. A second order Legendre polynomial consistent with a one-photon ionization process was used, and the images are shown in Fig. 7.2. along with the extracted anisotropy parameter (β) across the energy range 12.9–15.0 eV. For any dissociation, the anisotropy parameters range from +2 for when the axis of the bond that is broken is parallel to the electric vector of the VUV radiation, to −1 when the axis of the bond that is broken is perpendicular to that electric vector. The lobes of the distributions are perpendicular to the electric vector of the synchrotron light confirming the anisotropic distribution over the first photoelectron band corresponding to the ground electronic state of $SF_5CF_3^+$ [8]. However, the minimum value of the β parameter is −1.3 at 13.6 eV which outside the above limits. Nonetheless, the presence of anisotropy suggests that a revision of the *DIE* of SF_5CF_3 and the *AIE* of CF_3 determined by Chim et al. may be warranted. The radius of Argon at 15.95 eV corresponds to $3/2\ k_BT$, or 0.038 eV and was used to calibrate the KER, to determine the TKER values which shown in Fig. 7.3. These TKER values suggest a *DIE* of ca. 11.0 eV which is much lower than 12.9 eV determined by Chim et al. Further ion images and TKER calculated for CF_4, CCl_4 and SF_6 can be found in Appendix D. Whilst these results, obtained in a simple ion imaging experiment, hint at the scope of the future i^2PEPICO experiment, they are not rigorous enough on their own to draw firm conclusions other than the clear presence of anisotropy in dissociation from the ground states of SF_6^+ and $SF_5CF_3^+$.

Since the completion of my experimental work, a secondary imaging detector has been incorporated into the original iPEPICO apparatus at the Swiss Light Source together with a pulsed molecular beam source. Preliminary studies of

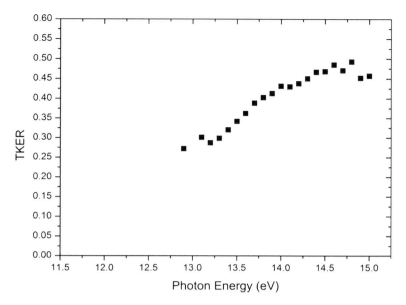

Fig. 7.3 Total kinetic energy release in CF_3^+ across the energy range 12.7–15.0 eV which corresponds to the first photoelectron band

the kinetic energy release in the photo fragments of the dissociation of $CF_4^+ \rightarrow CF_3^+ + F$, by Bodi et al. [11], allude to what contributions the different mechanisms of indirect (via Rydberg states) and direct ionization make to the overall photoelectron signal over the energy range of the first ion state. They also raised the possibility of imaging both cation and anion fragments produced through ion-pair formation. Such experiments are difficult because the overlap between the nuclear wave function of the molecular ground state and that of the ion-pair state is generally weak. However, with the higher flux and number densities in the ionization region delivered with an updated monochromator configuration, the i^{2}PEPICO could be used successfully with the tuneable synchrotron light to make such studies feasible [11]. An i^{2}PEPICO experiment has also recently been constructed by Tang et al. which successfully images both the threshold electron and corresponding ion in coincidence [3].

References

1. Bodi, A., Stevens, W. R., & Baer, T. J. (2011). *Journal of Physical Chemistry A, 115*, 726–734.
2. Bodi, A., Johnson, M., Gerber, T., Gengeliczki, Z., Sztáray, B., & Baer, T. (2009). *Review of Scientific Instruments, 80*, 034101.
3. Tang, X., Zhou, X., Wu, M., Gao, Z., Liu, S., Liu, F., et al. (2013). *Chemical Physics, 138*, 094306–094315.

4. Ashfold, M. N. R., Nahler, N. H., Orr-Ewing, A. J., Vieuxmaire, O. P. J., Toomes, R. L., Kitsopoulos, T. N., et al. (2005). *Physical Chemistry Chemical Physics, 8*, 26–53.
5. Hockett, P., Staniforth, M., Reid, K. L., & Townsend, D. (2009). *Physical Review Letters, 102*, 253002–253006.
6. Garcia, G. A., Nahon, L., & Powis, I. (2004). *Review of Scientific Instrumentation, 75*, 4989–4996.
7. Holland, M. P., West, J. B., Parr, A. C., Ederer, D. L., Stockbauer, R., Buff, R. D., et al. (1983). *Chemical Physics, 78*, 124–130.
8. Chim, R. Y., Kennedy, R. A., Tuckett, R. P., Zhou, W., Jarvis, G. K., Collins, D. J., et al. (2001). *Physical Chemistry A, 105*, 8403–8412.
9. Kinugawa, T., Hikosaka, Y., Hodgekins, A. M., & Eland, J. H. D. J. (2002). *Mass Spectrometry, 37*, 854–857.
10. Manzhos, S., & Loock, H.-P. (2003). *Computer Physics Communications, 154*, 76–87.
11. Bodi, A., Hemberger, P., Gerber, T., & Sztáray, B. (2012). *Review of Scientific Instrumentation, 83*, 083105–083113.

Appendix A

Chapter 4

Tables of 0 K total energies including the zero point energies which were calculated using a variety of ab initio techniques and the results of varying the weighting of different input parameters into the network, and used in Chap. 4, are presented below (Tables A.1, A.2, A.3, A.4, A.5, A.6).

Table A.1 Total 0 K energies including the zero point energy calculated with G3B3 and W1methods, with total 0 K energies calculated by Csontos

Species	G3B3 (0K) H	W1 (0K) H	Csontos neutrals $\Delta_f H^\circ_{0K}$ (H)
CHCl$_3$	-1418.833349	-1423.181789	-1423.643557
CH$_2$Cl$_2$	-959.374641	-962.281806	-962.591305
CH$_3$Cl	-499.915194	-501.379808	-501.537503
CCl$_4$	-1878.288639	-1884.07676	-1884.690439
CH$_2$Cl$^+$	-498.938772	-500.401326	
CHCl$_2^+$	-958.419567	-961.324125	
CCl$_3^+$	-1417.889685	-1422.235445	
CH$_3^+$	-39.431362	-39.452772	
CH$_3$F	-139.652064	-139.800335	-139.820583
CH$_2$F$_2$	-238.866356	-239.142222	-239.175648
CHF$_3$	-338.092025	-338.495796	-338.542544
CF$_4$	-437.314451	-437.846106	-437.907502
CH$_4$	-40.45827	-40.479069	
CH$_2$F$^+$	-138.659846	-138.809009	
CHF$_2^+$	-237.88441	-238.161657	
CF$_3^+$	-337.088561	-337.493773	
CF$_3$Cl	-797.552468	-799.39851	-799.597683
CF$_2$Cl$_2$	-1157.79383	-1160.953894	-1161.291212
CFCl$_3$	-1518.038981	-1522.513079	-1522.988716
CHClF$_2$	-698.333114	-700.052131	-700.23742
CCl$_2$F$^+$	-1057.624228	-1060.656081	
CClF$_2^+$	-697.358671	-699.076885	
CBrClF$_2$	-3271.296242		

Species highlighted in yellow are the values fitted in this work to derive new enthalpies offormation

Table A.2 Enthalpies of formation from a relaxed fit with no error minimization and the weighted (with values given in text) relaxed fit where all values except anchor values are included in the fit

Species	no weight relaxed fit	$\Delta_f H°_{0K}$ kJ mol^{-1} fit	weighted relaxed fit	$\Delta_f H°_{0K}$ kJ mol^{-1} fit
$CHCl_3$	anchor	-98.40	anchor	-98.40
CH_2Cl_2		-87.93		-87.89
CH_3Cl		-75.01		-75.62
CCl_4		-94.77		-94.82
CH_2Cl^+		962.15		960.79
$CHCl_2^+$		893.82		890.32
CCl_3^+		857.21		850.06
CH_3^+	anchor	1099.35	anchor	1099.35
CH_3F		-226.87		-227.44
CH_2F_2		-441.45		-442.34
CHF_3		-687.16		-687.75
CF_4	anchor	-927.80	anchor	-927.80
CH_4		-66.56		-66.56
CH_2F^+		846.14		845.38
CHF_2^+		602.57		601.72
CF_3^+	anchor	413.40	anchor	413.40
CF_3Cl		-704.88		-704.68
CF_2Cl_2		-490.75		-490.36
$CFCl_3$		-287.06		-286.47
$CHClF_2$		-476.09		-475.70
CCl_2F^+		706.14		701.37
$CClF_2^+$		554.63		552.24
$CBrClF_2$		-444.14		-446.52

Values highlighted in yellow are the final fit parameters, those highlighted in blue are the fixed anchor values

Table A.3 Enthalpies of formation with no weightings applied derived from the fit, and the original values, and the same again but with the experimentally derived enthalpies of formation weighted

No weighting	$\Delta_f H^\circ{}_{0K}$ (kJ mol^{-1}) fit	$\Delta_f H^\circ{}_{0K}$ (kJ mol^{-1})	Δ	iPEPICO weighting = 100 $\Delta_f H^\circ{}_{0K}$ fit	$\Delta_f H^\circ{}_{0K}$ (kJ mol^{-1})	Δ
CHCl$_3$	−98.40	−98.40		−98.40	−98.40	
CH$_2$Cl$_2$	−87.93	−88.66	0.73	−87.91	−88.66	0.75
CH$_3$Cl	−75.01	−74.30	−0.71	−75.65	−74.30	−1.35
CCl$_4$	−94.77	−88.70		−94.81	−88.70	
CH$_2$Cl$^+$	962.15	961.10	1.05	960.77	961.10	−0.33
CHCl$_2{}^+$	893.82	891.70		890.32	891.70	
CCl$_3{}^+$	857.21	834.60		850.10	834.60	
CH$_3{}^+$	1099.35	1099.35		1099.35	1099.35	
CH$_3$F	−226.87	−228.50	1.63	−227.48	−228.50	1.02
CH$_2$F$_2$	−441.45	−442.60	1.15	−442.37	−442.60	0.23
CHF$_3$	−687.16	−687.70	0.54	−687.76	−687.70	−0.06
CF$_4$	−927.80	−927.80		−927.80	−927.80	
CH$_4$	−66.56	−66.56		−66.56	−66.56	
CH$_2$F$^+$	846.14	836.50		845.34	836.50	
CHF$_2{}^+$	602.57	616.58		601.69	616.58	
CF$_3{}^+$	413.40	413.40		413.40	413.40	
CF$_3$Cl	−704.88	−703.40		−704.69	−703.40	
CF$_2$Cl$_2$	−490.75	−487.90		−490.37	−487.90	
CFCl$_3$	−287.06	−282.70		−286.48	−282.70	
CHClF$_2$	−476.09	−475.70	−0.39	−475.74	−475.70	−0.04
CCl$_2$F$^+$	706.14	712.71		701.40	712.71	
CClF$_2{}^+$	554.63	526.00		552.26	526.00	
CBrClF$_2$	−444.14	−423.80		−446.50	−423.80	
		Absolute x	0.57		Absolute x	0.03
		$\sigma\Delta$	0.84		$\sigma\Delta$	0.77
		Maximum deviation of $\sigma\Delta$	1.63		Maximum deviation of $\sigma\Delta$	1.35

Appendix A

Table A.4 Final enthalpies of formation derived from fitting only those highlighted in yellow

Species	Derived final values	$\Delta_f H°$ 0K kJ mol^{-1} fit	$\Delta_f H°$ 298K kJ mol^{-1}	$H_{298K}-H_{0K}$ kJ mol^{-1}
CHCl$_3$	anchor	-98.4	-103.4	14.1^a
CH$_2$Cl$_2$		-88.7	-95.6	11.8^a
CH$_3$Cl		-74.3	-82.2	10.4^a
CCl$_4$		-94.0	-96.4	17.1^b
CH$_2$Cl$^+$		961.1	957.1	10.1^b
CHCl$_2^+$		890.3	887.2	11.3^b
CCl$_3^+$		849.8	848.3	13.3^b
CH$_3^+$	anchor	1099.4	1095.6	10.0^b
CH$_3$F		-228.5	-236.6	10.1^a
CH$_2$F$_2$		-442.6	-450.3	10.6^a
CHF$_3$		-687.7	-694.7	11.5^a
CF$_4$	anchor	-927.8	-933.7	12.8^b
CH$_4$		-66.6	-74.6	10.0^b
CH$_2$F$^+$		844.4	840.4	10.0^b
CHF$_2^+$		601.6	598.4	11.0^b
CF$_3^+$	anchor	413.4	410.2	11.1^b
CF$_3$Cl		-702.1	-707.3	13.7^a
CF$_2$Cl$_2$		-487.8	-492.1	14.8^a
CFCl$_3$		-285.2	-288.6	15.9^a
CHClF$_2$		-475.7	-482.1	12.3^a
CCl$_2$F$^+$		701.2	699.0	12.5^b
CClF$_2^+$		552.2	549.5	11.8^b
CBrClF$_2$		-446.6	-457.6	15.7^b

^{a} from Csontos,[5] ^{b} from W1 calculations

Anchor values and non-highlighted species were not altered, together with the enthalpy correction

Table A.5 Enthalpies of formation derived with the experimental data at different weightings

iPEPICO weighting =	$\Delta_f H°_{0K}$ (kJ mol^{-1})					
	1.0	10.0	100.0	1000.0	10000.0	variance
CHCl$_3$	-98.4	-98.4	-98.4	-98.4	-98.4	0.0
CH$_2$Cl$_2$	-87.9	-88.1	-87.9	-88.1	-88.1	0.2
CH$_3$Cl	-75.0	-75.5	-75.6	-75.9	-75.9	0.9
CCl$_4$	-94.8	-94.6	-94.8	-94.8	-94.8	0.2
CH$_2$Cl$^+$	962.2	960.8	960.8	960.5	960.5	1.6
CHCl$_2^+$	893.8	890.3	890.3	890.3	890.3	3.5
CCl$_3^+$	857.2	850.0	850.1	853.4	853.4	7.2
CH$_3^+$	1099.4	1099.4	1099.4	1099.4	1099.4	0.0
CH$_3$F	-226.9	-227.4	-227.5	-227.3	-227.3	0.6
CH$_2$F$_2$	-441.5	-442.2	-442.4	-442.2	-442.2	0.9
CHF$_3$	-687.2	-687.7	-687.8	-687.7	-687.7	0.6
CF$_4$	-927.8	-927.8	-927.8	-927.8	-927.8	0.0
CH$_4$	-66.6	-66.6	-66.6	-66.6	-66.6	0.0
CH$_2$F$^+$	846.1	845.5	845.3	845.5	845.5	0.8
CHF$_2^+$	602.6	601.8	601.7	601.8	601.8	0.9
CF$_3^+$	413.4	413.4	413.4	413.4	413.4	0.0
CClF$_3$	-704.9	-704.8	-704.7	-704.6	-704.6	0.3
CF$_2$Cl$_2$	-490.8	-490.5	-490.4	-490.3	-490.3	0.5
CFCl$_3$	-287.1	-286.7	-286.5	-286.5	-286.5	0.6
CHClF$_2$	-476.1	-475.8	-475.7	-475.6	-475.6	0.5
CCl$_2$F$^+$	706.1	701.4	701.4	703.5	703.5	4.8
CClF$_2^+$	554.6	552.2	552.3	553.3	553.2	2.4
CBrClF$_2$	-444.1	-446.5	-446.5	-445.5	-445.5	2.4

The final values used with a weighting of 100 for the experimental onsets are highlighted in yellow

Appendix A

Table A.6 Ab initio calculated enthalpies of formation, values from the Csontos study and an example of how the values change, fit, is given in the last column

	G3B3 $\Delta_f H^{\ominus}$ (0 K) Hartrees	CBS-QB3 $\Delta_f H^{\ominus}$ (0 K) Hartrees	CBS-APNO $\Delta_f H^{\ominus}$ (0 K) Hartrees	W1 $\Delta_f H^{\ominus}$ (0 K) Hartrees	Csontos $\Delta_f H^{\ominus}$ neutrals Hartrees
$CHCl_3$	−1418.833349	−1417.870203		−1423.181789	−1423.643557
CH_2Cl_2	−959.374641	−958.716555		−962.281806	−962.591305
CH_3Cl	−499.915194	−499.562049		−501.379808	−501.537503
CCl_4	−1878.29	−1877.02		−1884.08	−1884.6904
CH_2Cl^+	−498.938772	−498.585752		−500.401	
$CHCl_2^+$	−958.419567	−957.760866		−961.324	
CCl_3^+	−1417.889685	−1416.925283		−1422.24	
CH_3^+	−39.431362	−39.384653	−39.4413	−39.4528	
Cl_2	−920.074	−919.461		−922.958	−923.262
F_2	−199.428914	−199.512588	−199.513	−199.682	−199.711
Br_2	−5147.110006				
Cl_2 bond energy (H)		0.091106469			
F_2 bond energy (H)		0.058884031			
Br_2 bond energy (H)		0.072806714			
H atom	−0.501087	−0.499946	−0.49995	−0.49999	

	G3B3 (0 K) Hartrees	CBS-QB3 (0 K) Hartrees	CBS-APNO (0 K) Hartrees	W1 (0 K) Hartrees	Csontos neutrals
CH_3F	−139.652064		−139.704	−139.8	−139.821
CH_2F_2	−238.866356		−238.961	−239.142	−239.176
CHF_3	−338.092025		−338.23	−338.496	−338.543
CF_4	−437.314451		−437.495	−437.846	−437.908
CH_4	−40.45827	−40.41	−40.4689	−40.4791	
CH_2F^+	−138.659846		−138.713	−138.809	
CHF_2^+	−237.88441		−237.981	−238.162	
CF_3^+	−337.088561		−337.228	−337.494	

(continued)

Table A.6 (continued)

	G3B3 (0 K) Hartrees	CBS-QB3 (0 K) Hartrees	CBS-APNO (0 K) Hartrees	W1 (0 K) Hartrees	Csontos neutrals
$CClF_3$	−797.552468	−797.07839		−799.399	−799.598
CF_2Cl_2	−1157.79383			−1160.95	−1161.29
$CFCl_3$	−1518.038981	−1517.035241		−1522.51	−1522.99
$CHClF_2$	−698.333114			−700.052	−700.237
$CHCl_2F$	−1058.580311			−1061.61	−1061.94
$CHClF^+$	−598.150912	−597.757429		−599.741	
CCl_2F^+	−1057.624228	−1056.925016		−1060.66	
$CClF_2^+$	−697.358671	−696.924563		−699.077	

	G3B3 (0 K) Hartrees	CBS-QB3 (0 K) Hartrees	CBS-APNO (0 K) Hartrees	W1 (0 K) Hartrees	Csontos neutrals
$CBrClF_2$	−3271.296242				
CBr_2F_2	−5384.798755				
CBr_4	−10332.30704				
CF_3Br	−2911.054215				
CH_3Br	−2613.42163				

Appendix B

Chapters 4 and 5

Input data for modelling the breakdown curves in Chap.4, calculated with B3LYP/ 6-311 ++G** basis set and level of theory using Gaussian 03 (Fig. B.1).

CH_3Cl

Neutral frequencies/cm^{-1}	710, 1025, 1025, 1377, 1483, 1483, 3072, 3167, 3167
Neutral rotational constants/Hz	157502390000, 13096080000, 13096020000
Ion frequencies/cm^{-1}	173, 634, 825, 1158, 1226, 1414, 2778, 3061
Transition state 1 frequencies/cm^{-1}	485, 499, 1288, 1404, 1532, 3062, 3261, 3268

CH_2Cl_2

Neutral frequencies/cm-1	283, 704, 727, 914, 1196, 1321, 1489, 3144, 3223
Neutral rotational constants/Hz	32180670000, 3194240000, 2961290000
Ion frequencies/cm-1	262, 600, 743, 963, 1110, 1274, 1470, 3155, 3265
Transition state 1 frequencies/cm-1	243, 702, 892, 1054, 1134, 1452, 3125, 3266

$CHCl_3$

Neutral frequencies/cm^{-1}	260, 260, 367, 667, 737, 737, 264, 1264, 3203
Neutral rotational constants/Hz	3197290000, 3197290000, 1657110000
Ion frequencies/cm^{-1}	149, 149, 354, 647, 647, 677, 1229, 1229, 3171
Transition state 1 frequencies/cm^{-1}	146, 287, 350, 739, 931, 976, 1291, 3193

CH$_3$F

Neutral frequencies/cm^{-1}	1092, 1204, 1204, 1523, 1523, 1531, 3038, 3112, 3112
Neutral rotational constants/Hz	156698870000, 25583990000, 25583990000
Ion frequencies/cm^{-1}	750, 964, 994, 1071, 1292, 1451, 2166, 2532, 3181
Transition state 1 frequencies/cm^{-1}	194, 252, 1214, 1232, 1410, 1552, 3049, 3222

CH$_2$F$_2$

Neutral frequencies/cm^{-1}	520, 1058, 1096, 1169, 1266, 1449, 1520, 3059, 3134
Neutral rotational constants/Hz	490631000000, 104471300000, 91377600000
Ion frequencies/cm^{-1}	589, 635, 993, 1055, 1102, 1227, 1381, 2142, 2494
Transition state 1 frequencies/cm^{-1}	154, 178, 660, 1019, 1302, 1347, 1604, 3151

CBrClF$_2$

Neutral frequencies/cm^{-1}	213, 290, 325, 404, 433, 639, 839, 1071, 1125
Neutral rotational constants/Hz	3813310000, 1649540000, 1443260000
Ion frequencies/cm^{-1}	171, 252, 265, 390, 433, 628, 678, 1164, 1314
Transition state 1 frequencies/cm^{-1}	162, 217, 402, 472, 549, 729, 1233, 1372

CHClF$_2$

Neutral frequencies/cm^{-1}	359, 400, 594, 778, 1109, 1129, 1319, 1374, 3130
Neutral rotational constants/Hz	10190040000, 4752230000, 3446910000
Ion frequencies/cm^{-1}	127, 197, 389, 641, 1034, 1287, 1296, 1532, 3183
Transition state 1 frequencies/cm^{-1}	54, 116, 662, 1012, 1304, 1352, 1602, 3155

Input data for modelling the breakdown curves in Chap. 5, calculated with B3LYP/6-311 + G** basis set and level of theory using Gaussian 03.

Monofluorethene

Neutral frequencies/cm^{-1}	484, 733, 871, 952, 977, 1191, 1350, 1441, 1742, 3192, 3216, 3285
Neutral rotational constants/Hz	65687710000, 10529410000, 9074770000
Ion frequencies/cm^{-1}	399, 486, 873, 989, 1013, 1253, 1333, 1469, 1587, 3173, 3216, 3293
Transition state 1 frequencies/cm^{-1}	341, 540, 718, 744, 777, 846, 1144, 1600, 1671, 3206, 3260
Transition state 2 frequencies/cm^{-1}	148, 246, 363, 389, 808, 824, 1069, 1279, 2103, 3031, 3141
Transition state 1 = HF loss, transition state 2 = H loss	

Appendix B

1,1-difluorethene

Neutral frequencies/cm^{-1}	430, 495, 572, 725, 868, 919, 953, 1260, 1451, 1763, 3194, 3293
Neutral rotational constants/Hz	10245430000, 10121060000, 5091430000
Ion frequencies/cm^{-1}	373, 417, 582, 627, 922, 958, 1005, 1428, 1515, 1578, 3140, 3268
Transition state 1 frequencies/cm^{-1}	87, 433, 496, 557, 573, 739, 1099, 1208, 1327, 2117, 3197
Transition state 2 frequencies/cm^{-1}	408, 465, 610, 918, 959, 1081, 1200, 1444, 1530, 3122, 3260
Transition state 1 = HF loss, transition state 2 = F loss	

Trifluorethene

Neutral frequencies/cm^{-1}	226, 306, 480, 556, 620, 775, 946, 1191, 1298, 1395, 1851, 3270
Neutral rotational constants/Hz	10517160000, 3832250000, 2808780000
Ion frequencies/cm^{-1}	221, 234, 486, 567, 639, 798, 959, 1287, 1415, 1573, 1667, 3234
Transition state 1 frequencies/cm^{-1}	35, 74, 112, 197, 621, 723, 1234, 1281, 1468, 1506, 3213
Transition state 2 frequencies/cm^{-1}	35, 74, 112, 197, 621, 723, 1234, 1281, 1468, 1506, 3213
Transition state 3 frequencies/cm^{-1}	15, 54, 75, 123, 146, 649, 1094, 1256, 1282, 1497, 2787
Transition state 1 = CF loss, transition state 2 = CHF loss, transition state 3 = CF$_2$ loss	

Tetrafluorethene

Neutral frequencies/cm^{-1}	193, 200, 391, 415, 480, 544, 545, 798, 1201, 1361, 1367, 1928
Neutral rotational constants/Hz	5411200000, 3225030000, 2020710000
Ion frequencies/cm^{-1}	137, 208, 400, 426, 530, 548, 591, 834, 1296, 1568, 1569, 1725
Transition state 1 frequencies Regime 1/cm^{-1}	19, 43, 46, 110, 472, 540, 541, 912, 1473, 1513, 1514
Transition state 2 frequencies Regime 1/cm^{-1}	30, 65, 72, 152, 110, 472, 540, 541, 912, 1473, 1513, 1514
Transition state 3 frequencies regime 1/cm^{-1}	19, 45, 50, 98, 116, 607, 663, 1149, 1321, 1388, 1395
Neutral frequencies regime 2/cm^{-1}	193, 200, 391, 415, 480, 544, 545, 798, 1201, 1361, 1367, 1928
Neutral rotational constants regime 2/Hz	5411200000, 3225030000, 2020710000
Ion frequencies regime 2/cm^{-1}	137, 208, 400, 426, 530, 548, 591, 834, 1296, 1568, 1569, 1725

(continued)

(continued)

Transition state 1 frequencies regime 2/cm^{-1}	25, 40, 60, 200, 220, 400, 426, 530, 548, 591, 834, 1296, 1568, 1569, 1725
Transition state 2 frequencies regime 2/cm^{-1}	25, 40, 60, 200, 220, 400, 426, 530, 548, 591, 834, 1296, 1568, 1569, 1725

Regime 1; Transition state 1 = CF loss, transition state 2 = CF$_3$ loss, transition state 3 = CF$_2$ loss
Regime 2: Transition state 1 = F loss, transition state 2 = CF$_2$ + F loss

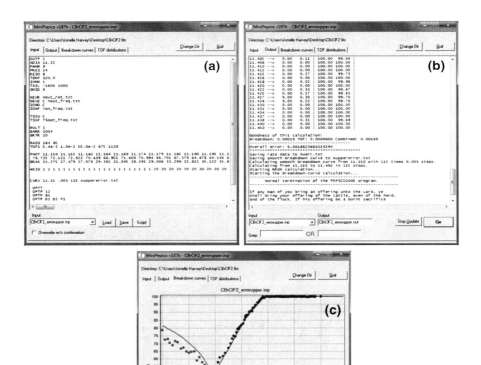

Fig. B.1 Screenshots of breakdown curve modelling program as described in Chap. 3. **a** Input experiment data and experimental paramters **b** Output data showing successful completion of the calculation **c** Output modelled curves (*line*) and experimental data points (*solid shapes*)

Appendix C

Threshold Photoelectron Spectra of Halogenated Methanes Recorded with the iPEPICO Apparatus at the VUV Beamline at the Swiss Light Source

Acknowledgement is given to Dr. Andras Bodi, Dr. Melanie Johnson, Nicola Rogers, Dr. Matthew Simpson, and Prof. Richard Tuckett for assistance with data collection. All spectra are normalized to sample pressure and photon flux (Figs. C.1, C.2, C.3, C.4, C.5).

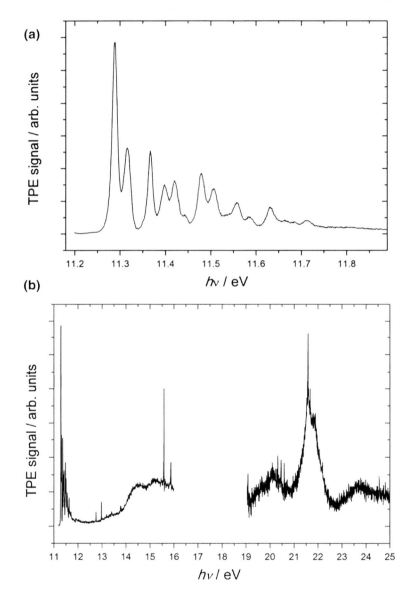

Fig. C.1 a TPES of the ground electronic state of CH$_3$Cl recorded in April 2009, from 11.2–11.9 eV with a step size of 0.001 eV and an integration time of 40 s per point **b** TPES of the ground electronic state and the first and second excited states of CH$_3$Cl (*first portion*). Recorded from 11.2–11.9 eV, with a step size of 0.001 eV and an integration time of 40 s per point. Recorded from 11.9–16.0 eV with a step size of 0.005 eV and an integration time of 40 s per point. The *second portion* is recorded from 19.0–25.0 eV, with a step size of 0.005 eV and an integration time of 40 s per point

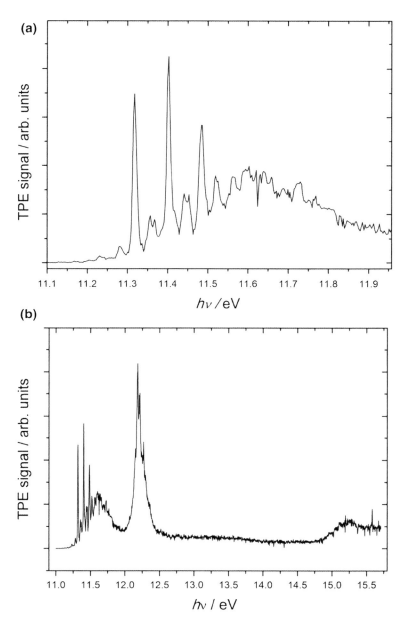

Fig. C.2 **a** TPES of the ground and first electronic state of CH_2Cl_2 recorded in April 2009, from 11.05–11.30 eV with a step size of 0.003 eV and an integration time of 40 s per point **b** TPES of the ground and first electronic state and the first and second excited states of CH_2Cl_2 Recorded from 11.05–11.30 eV, with a step size of 0.003 eV and an integration time of 40 s per point. Recorded from 11.30–15.7 eV with a step size of 0.003 eV and an integration time of 40 s per point

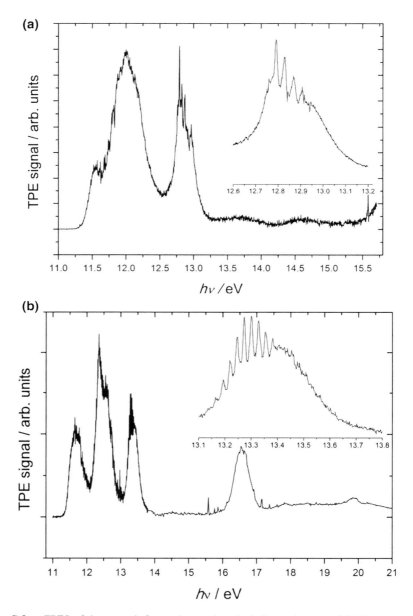

Fig. C.3 a TPES of the ground, first and second excited electronic state of CHCl$_3$ recorded in March 2009, from 11.00–15.71 eV with a step size of 0.003 eV and an integration time of 26 s per point. Inset shows as more detailed spectrum from 12.60–13.20 with a step size of 0.001 eV and an integration time of 44 s **b** TPES of the ground, first second, third and fourth excited electronic state and the first and second excited states of CCl$_4$ Recorded from 11.00–14.00 eV, with a step size of 0.003 eV and an integration time of 40 s per point. Recorded from 14.0–21 eV with a step size of 0.02 eV and an integration time of 60 s per point. Inset, recorded 13.10–13.80 eV, 40 s integration time, 0.002 step size

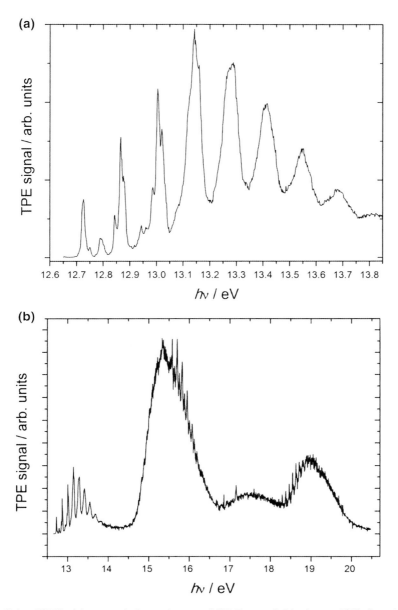

Fig. C.4 **a** TPES of the ground electronic state of CH_2F_2 recorded in August 2010, from 12.65–13.85 eV, with an integration time of 90 s and a step size of 0.002 eV **b** TPES of the ground and first six excited electronic states of CH_2F_2 recorded in August 2010. 12.65–13.85 eV, with an integration time of 30 s and a step size of 0.002 eV. 14.60–16.40 eV, with an integration time of 30 s and a step size of 0.002 eV, 16.40–18.00 eV, with an integration time of 30 s and a step size of 0.006 eV, 18.00–19.80 eV, with an integration time of 30 s and a step size of 0.002 eV

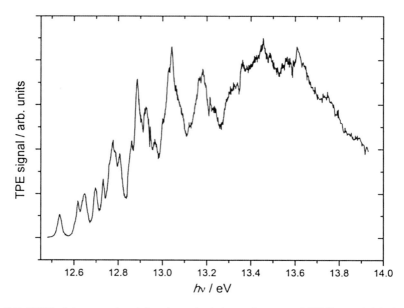

Fig. C.5 TPES of the ground and first six excited electronic states of CH$_3$F recorded in August 2010. 12.48–13.60 eV with a step size of 0.001 eV, integration time 35. 13.60–13.93 eV with a step size of 0.002 eV and integration tome of 35 s

Appendix D

Ion Images

Raw ion images of CF_3^+, SF_5^+ and CCl_4^+ as captured by the Roentdek DLD40 detector, using a thermal gas source. Contamination in the images by background signal originating from O_2^+ and N_2^+ account for the strong central spot in each image. The total kinetic energy release is plotted for CF_3^+, and CCl_4^+ across the energy ranges of the first photoelectron bands. The raw images were reconstructed with the onion peeling algorithm. The radius of the image is proportional to the kinetic energy release. Argon was used to calibrate the images as once the Ar^+ formed, and excess energy is carried away by the electron, so the radius of the Argon image is equal to to 3/2 $k_B T$ or 0.038 eV. All ions were extracted with fields of 80 V cm^{-1} (Figs. D.1, D.2, D.3, D.4, D.5).

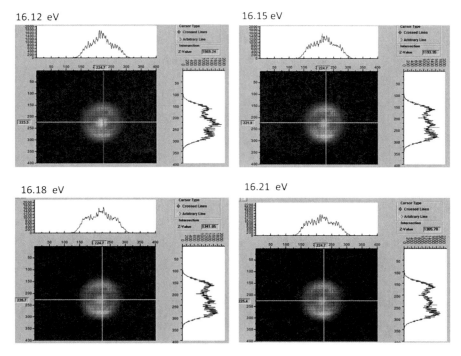

Fig. D.1 Screenshots of ion images of CF_3^+ from the reaction $CF_4 + h\nu \rightarrow CF_3 + F + e^-$ as captured by the Roentdek DLD40 detector. Side graphs show cross sections through the distributions through the mid-points

Appendix D

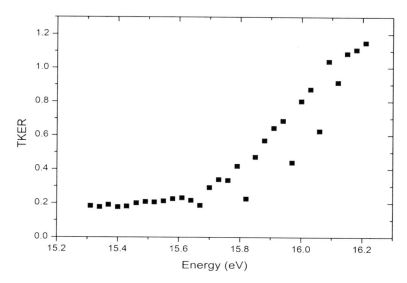

Fig. D.2 Kinetic energy release of CF_3^+ energy release of CF_3^+ from the reaction $CF_4 + h\nu \rightarrow CF_3 + F + e^-$. The 2-dimensional image was reconstructed using with the onion peeling algorithm, see Chap. 7. The radius of the image is proportional to the kinetic energy release. The CF_3^+ images were calibrated using the radius of the Argon image at 15.71 eV which is equal to $3/2\ k_B T$ or 0.038 eV

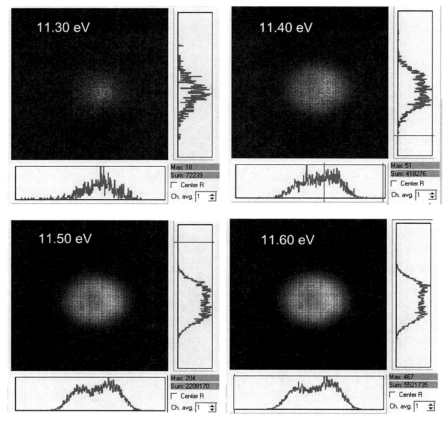

Fig. D.3 Screenshots of ion images of CCl_4^+ from the reaction $CCl_4 + h\nu \rightarrow CCl_3^+ + Cl + e^-$ as captured by the Roentdek DLD40 detector. Side graphs show cross sections through the distributions through the mid-points

Appendix D

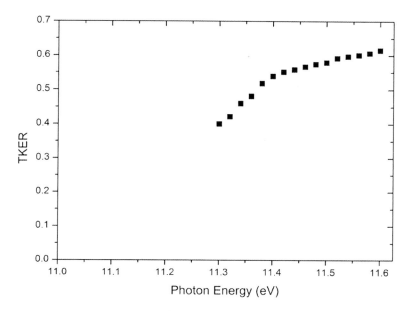

Fig. D.4 Kinetic energy release of CCl_3^+ from the reaction $CCl_4 + h\nu \rightarrow CCl_3^+ + Cl + e^-$. The 2-dimensional image was reconstructed using with the onion peeling algorithm, see Chap. 7. The radius of the image is proportional to the kinetic energy release. The CF_3^+ images were calibrated using the radius of the Argon image at 15.71 eV which is equal to 3/2 $k_B T$ or 0.038 eV

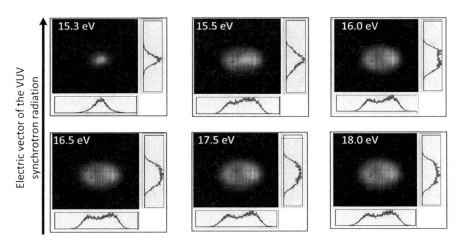

Fig. D.5 Screenshots of ion images of SF_5^+ from the reaction $SF_6 + h\nu \rightarrow SF_5 + F + e^-$ as captured by the Roentdek DLD40 detector, showing clear anisotropy, perpendicular to the electric vector of the VUV light. Side graphs show cross sections through the distributions through the mid-points

Printed by Publishers' Graphics LLC
CAMZ131112.15.19.120